John Curwen

History of the Association of Medical Superintendents of American

Institutions for the Insane

From 1844 to 1884

John Curwen

History of the Association of Medical Superintendents of American Institutions for the Insane
From 1844 to 1884

ISBN/EAN: 9783337375539

Printed in Europe, USA, Canada, Australia, Japan

Cover: Foto ©berggeist007 / pixelio.de

More available books at **www.hansebooks.com**

HISTORY OF THE ASSOCIATION

OF

MEDICAL SUPERINTENDENTS

OF

American Institutions for the Insane,

FROM 1844 TO 1884, INCLUSIVE,

WITH

A LIST OF THE DIFFERENT HOSPITALS FOR THE
INSANE, AND THE NAMES AND DATES OF AP-
POINTMENT AND RESIGNATION OF THE
MEDICAL SUPERINTENDENTS.

——o——

Compiled from the Records of the Association

BY

JOHN CURWEN, M. D.

SECRETARY OF THE ASSOCIATION.

——o——

1885.

WARREN, PA.:
E. COWAN & CO., PRINTERS.

ERRATA.

———

Page   18, line   5, read " or" for " and."

Page   46, line 13, read " decease" for " disease."

Page   62, line 21, read " Joseph" for " John."

Page   79, line 14, read " of" for " in."

Page 116, line   4, read " such" for " each."

Page 121, line   2, read " 1877" for " 1874."

HISTORY OF THE ASSOCIATION.

———o———

To whom the honor of first proposing the formation of this Association is due, may be fully learned from a letter written to the compiler of this sketch, by one of the oldest and most distinguished physicians in connection with a hospital for the insane (Dr. Francis T. Stribling).

"About the year 1844, Dr. Woodward, the able and distinguished pioneer in our specialty (then superintending the State Lunatic Hospital, at Worcester, Massachusetts), honored me with a visit. We, of course, conferred freely as to all that concerned the interests of the insane ; the organization, management, &c., of institutions for the benefit of this afflicted class was the topic on which we dwelt much. At that period, there were comparatively few hospitals for the insane in the United States, and as we had felt mutually benefited by our interchange of views, it was deemed but reasonable that the noble cause would be materially promoted by some arrangement to convene at stated periods, for consultation, &c., with those in charge of such institutions. If I forget not, it was understood between us that Dr. Woodward would, on his way home, have a personal interview with Dr. Kirkbride, and that they would, by mail, or othewise, confer with Dr. Awl, of Ohio, and probably some others. The result was a meeting of medical superintendents, at Philadelphia, and the organization of an association, which in my opinion, has secured to the American asylums for the insane, a reputation and a confidence which the heretofore boasted of institutions in Europe cannot afford to ' look down upon.' "

The first meeting of the Association was held at Jones' hotel in the city of Philadelphia, on the 16th day of October, 1844, at 10 o'clock, A. M.

The officers appointed were Dr. Samuel B. Woodward, President ; Dr. Samuel White, Vice President, and Dr. Thomas S. Kirkbride, Secretary and Treasurer.

It was decided that the Convention should be styled, " The Convention of Medical Superintendents and Physicians of the Asylums and Hospitals for the Insane in the United States, and shall consist of the following individuals and of such other medical superintendents and physicians of asylums or hospitals for the insane as may hereafter be admitted by a vote of the majority of the members of the Convention," viz:

Dr. Samuel B. Woodward, of the Massachusetts State Lunatic Hospital, at Worcester.

Dr. Samuel White, of the Hudson Lunatic Asylum, Hudson, New York.

Dr. Isaac Ray, of the Maine Insane Hospital, at Augusta.

Dr. Luther V. Bell, of the McLean Asylum for the Insane, at Somerville, Massachusetts.

Dr. C. H. Stedman, of the Boston Lunatic Hopsital, Boston, Massachusetts.

Dr. John S. Butler, of the Connecticut Retreat for the Insane, at Hartford.

Dr. Amariah Brigham, of the New York State Lunatic Asylum, at Utica.

Dr. Pliny Earle, of the Bloomingdale Asylum, New York.

Dr. Thomas S. Kirkbride, of the Pennsylvania Hospital for the Insane, Philadelphia.

Dr. Wm. M. Awl, of the Ohio Lunatic Asylum, Columbus.

Dr. Francis T. Stribling, of the Western Lunatic Asylum of Virginia, at Staunton.

Dr. John M. Galt, of the Eastern Lunatic Asylum, at Williamsburg, Virginia.

Dr. Nehemiah Cutter, of the Pepperell Private Asylum, Pepperell, Massachusetts.

The committee appointed to prepare business for the Convention recommended the appointment of committees on fifteen different subjects ; which was adopted, and the subjects and the gentlemen selected to compose the committees were as follows :

1. On the Moral Treatment of Insanity—Drs. Brigham, Cutler and Stribling.
2. On the Medical Treatment of Insanity—Drs. Woodward, Awl and Bell.
3. On Restraint and Restraining Apparatus—Drs. Bell, Ray and Stedman.
4. On the Construction of Hospitals for the Insane—Drs. Awl, White, Bell, Butler, Galt and Ray.
5. On the Jurisprudence of Insanity—Drs. Ray, Stribling and Stedman.
6. On the Prevention of Suicide—Drs. Butler, Kirkbride and Earle.
7. On the Organization of Hospitals for the Insane and a Manual for Attendants—Drs. Kirkbride, Brigham and Galt.
8. On the Statistics of Insanity—Drs. Earle, Ray and Awl.
9. On the Support of the Pauper Insane—Drs. Stribling, Bell and Ray.
10. On Asylums for Idiots and Demented—Drs. Brigham, Awl and White.
11. On Chapels and Chaplains in Insane Hospitals—Drs. Butler, White and Stedman.
12. On Postmortem Examinations—Drs. Kirkbride, Stedman and Galt.
13. On Comparative Advantages of Treatment in Hospitals and in Private Practice—Drs. White, Ray and Butler.
14. On Asylums for Colored Persons—Drs. Galt, Awl and Stribling.
15. On Provisions for Insane Prisoners—Drs. Brigham, Bell and Earle.
16. On Causes and Prevention of Insanity—Drs. Stribling, Kirkbride and Brigham.

Reports were presented from nearly all of these committees, and discussions held on the general subjects mentioned, though the committees were all continued, with the expectation of presenting fuller reports at a subsequent meeting.

In a report on moral treatment, strong ground is taken in favor of schools in hospitals for certain classes of patients, as a means of mental occupation.

The following is of interest :

Resolved, That it is the unanimous sense of this Convention that the attempt to abandon entirely the use of all means of personal restraint is not sanctioned by the true interests of the insane.

Towards the close of the session, it was resolved that the title of the body in future shall be, " The Association of Medical Superintendents of American Institutions for the Insane," and that the medical superintendents of the various incorporated or other legally constituted institutions for the insane, now existing in the United States, or which may be commenced prior to the next meeting, be and they are hereby elected members of this Association.

It was also resolved that any member or members of this Association who may be in Europe at the time of the meeting of the convention of physicians to the institutions for the Insane of Great Britain, be authorized to represent this body at that meeting, and that the President and Secretary furnish the proper credentials.

The Association adjourned on October 19, to meet in Washington, D. C., on the second Monday in May, 1846.

Dr. Samuel White, the Vice President, died before the next meeting of the Association, and from a memoir prepared for a subsequent meeting, the following abstract is made :

Dr. White was born in Coventry, Connecticut, on February 23, 1777, and pursued the study of medicine and surgery with Dr. Philip Turner, of Norwich, Connecticut, a distinguished surgeon in the army of the revolution. He commenced his professional career at Hudson, New York, 1797, and married in 1799. His practice soon became extensive and he was often called especially as a surgeon to a great distance. In 1808 he was elected Professor of Obstetrics and Practical Surgery in the Berkshire Medical Institution, Pittsfield, Massachusetts, which situation, after giving two courses of lectures, he resigned.

Owing to the occurrence of insanity in his own family, by which his domestic enjoyments were interrupted, he was led to pay much attention to this disease, and in 1830 he established a private lunatic asylum at Hudson, which he successfully cone ducted. In 1843 he was elected President of the New York State Medical Society, and delivered an address on insanity, which presented one of the best synopsis of our knowledge of insanity, especially of its treatment, which has ever been published. In October, 1844, he attended the meeting of the Association, but soon after his health began to fail, and he died at Hudson, February 10, 1845.

In his personal appearance he was tall, though slender, his countenance grave and dignified, yet he was of a social disposition, and a man of pleasing address. He discharged the various duties of a long and active professional life with ability and in a truly Christian spirit.

Resolutions relative to Dr. White, offered by Dr. Brigham:

WHEREAS, Since the last meeting of this Association, Dr. Samuel White, of New York, the venerable and highly respected late Vice President of this Association, has died; therefore,

Resolved, That by the death of Dr. White, this Association and the medical profession have lost an esteemed and valued member, and the cause of humanity, a useful and active friend. Particularly have the friends of the insane reason to mourn his loss, as he has long been successfully engaged in relieving the sufferings of this afflicted class of his fellow beings, and, by his labors and his writings, essentially aided in improving their condition.

Resolved, That we deeply sympathize with the surviving members of his family, and recall at the present time the excellencies of his character, his useful precepts, and the worthy example he presented of a gentleman, physician and Christian, devoted to deeds of goodness, and whose long and active life was spent in promoting the welfare of his fellow men.

Resolved, That the Secretary of this Association present a copy of these resolutions to the nearest relative of the deceased.

————o————

At the second meeting of the Association, which was held in Washington, D. C., on May 11, 1846, the following gentlemen were present :

Dr. Samuel B. Woodward, of the Massachusetts State Lunatic Hospital, at Worcester.

Dr. James Bates, of the Maine Insane Hospital at Augusta.

Dr. Andrew McFarland, of the New Hampshire State Hospital, at Concord.

Dr. William H. Rockwell, of the Vermont State Hospital, at Brattleboro'.

Dr. Luther V. Bell, of the McLean Asylum for the Insane, Somerville, Massachusetts.

Dr. N. Cutter, of the Pepperell Private Asylum, Massachusetts.

Dr. C. H. Stedman, of the Boston Lunatic Hospital.

Dr. George Chandler, of the Massachusetts State Lunatic Hospital, at Worcester.

Dr. John S. Butler, of the Connecticut Retreat for the Insane, Hartford,

Dr. Pliny Earle, of the Bloomingdale Asylum, New York.

Dr. G. H. White, of the Hudson Lunatic Asylum, New York.

Dr. Thomas S. Kirkbride, of the Pennsylvania Hospital for the Insane, Philadelphia.

Dr. R. S. Steuart and Dr. John Fonerden, of the Maryland Hospital, Baltimore.

Dr. William H. Stokes, of the Mount Hope Asylum, Baltimore.

Dr. William M. Awl, of the Ohio Lunatic Asylum, Columbus.

Dr. John M. Galt, of the Eastern Asylum of Virginia, at Williamsburg.

Dr. J. W. Parker, of the South Carolina Hospital, Columbia.

Dr. Walter Telfer, of the Lunatic Hospital, Toronto, Canada.

Dr. James Macdonald, of the Private Institution, at Flushing, New York, and Dr. Amariah Brigham, of the New York State Lunatic Asylum, Utica, New York.

The resolution of the last meeting relative to members was amended so as to read : That the medical superintendents of the various incorporated or other legally constituted institutions for the insane now existing on this continent, or which may be commenced prior to the next meeting, and all those who have heretofore been medical superintendents and members of this Association, or who may be hereafter appointed to those stations be and they hereby are constituted members of the Association.

Dr. William M. Awl was elected Vice President, in place of Dr. Samuel White, deceased.

The following resolution was also adopted : That in future every regularly constituted institution for the insane on this continent may have a representative in this Association; that as heretofore, this shall be the medical superintendent where such officer exists ; but in those institutions where there is a different organization, it may be either of the regular medical officers who may find it most convenient to attend.

These two resolutions form the Constitution of the Association.

The committees appointed at the previous meeting generally presented reports which led to interesting discussions, and the following committees were appointed on the subjects named :

1. Treatment of Incurables—Dr. Macdonald.
2. Is there any relation between Phrenology and Insanity—Dr. Fonerden.
3. The Classification of Insanity—Dr. Earle.
4. The admission of visitors into the halls of patients—Dr. Ray.
5. Visits to and correspondence with patients by friends—Dr. Stokes.

6. The comparative value of the different kinds of manual labor for patients, and the best means of employment in winter—Dr. Rockwell.

7. The proper number of patients for one institution—Dr. Brigham.

8. The utility of night attendants, and the propriety of not locking the patients' doors at night—Dr. Chandler.

9. The advantages and disadvantages of cottages for wealthy patients adjacent to the hospitals for the insane—Dr. Kirkbride.

10. The relative value of different kinds of fuel for heating hospitals—Dr. Bates.

11. Insanity and the condition of the insane in the British Provinces—Dr. Telfer.

12. The nature and treatment of insanity produced by the use of intoxicating liquors—Dr. Stedman.

13. The relations of menstruation to insanity—Dr. Fonerden.

14. Under what circumstances can the insane of the poorer classes be properly treated with the greatest economy—Dr. McFarland.

15. The effects upon the insane of the use of tobacco—Dr. Cutter.

16. Reading, recreation and amusement for the insane—Dr. Galt.

17. On water closets in the wards and yards of hospitals for the insane—Dr. Bell.

18. On the construction and arrangement of institutions for the insane in southern climates—Dr. Parker.

The Association adjourned on May 14, to meet in New York on the second Monday of May, 1848.

Dr. Samuel Bayard Woodward was born at Torrington, Litchfield county, Connecticut, June 10, 1787 ; studied medicine with his father, an eminent physician, and at the age of twenty-one was licensed to practice medicine by the medical board of his native county, and soon after located himself in Weathersfield, Connecticut. When the Penitentiary was moved to Weathersfield, he was appointed physician, and held the position so long as he remained in Weathersfield.

He was one of the medical examiners of the medical school in New Haven, chosen by the State Medical Society. He was efficient in establishing the Retreat at Hartford. He issued circulars and made the arrangements to collect the funds. He was one of the medical visitors of that institution while he remained in the vicinity. He took credit to himself in having secured for it its present delightful location. His attention was called to this special department of the profession by the occurence of several cases of insanity in his own practice, and in that of his professional brethren, whose advisor he was. The difficulty of managing these cases in their private practice led Dr. Woodward, and his particular friend, Dr. Eli Todd, to take the first step towards the establishment of the retreat. He was appointed Superintendent of the State Lunatic Hospital, at Worcester, in September, 1832 ; went to Worcester in December following, and moved into the hospital as soon as rooms could be finished and furnished for the reception of his family. He retired on June 30, 1846, on account of failing health, and moved to Northampton, where he died quite suddenly on the evening of January 3, 1850.

The following interesting facts in his history may also be noticed :

At the age of four years he went, as was then the custom, to a pest house, and was inoculated with matter of small-pox. He had the disease quite severely. What was remarkable in his case was the fact of his having small-pox twice afterwards.

At about thirty years old, he was severely sick with a low grade of fever, and was delirious ; during which he refused to take food, under the impression that his children were given him to eat, and afterwards from thinking it wrong to eat, as there was not food enough on the earth to support the inhabitants. He said he was induced to take food by a stratagem of his father, who gave him water, saying to him that the Lord had made a great supply of water for all things, which seemed to him reasonable. When it was dark his father added milk to the water. In this way he was sustained until the delusion passed off. In that sickness he labored under a disease of the organ of the vision.

Those in his sick chamber seemed to move with great velocity. Those coming toward him appeared to be coming so fast that they would certainly dash against him. He requested them to move slowly. This fever left him with an enlarged leg, the veins of which were varicose. An ulcer came upon the ankle of this leg, which was open most of the time.

He was elected in 1830 to a seat in the Senate of Connecticut, and on taking up his residence in Massachusetts, was commissioned by the Governor as a justice of the peace, which was renewed at the end of seven years.

His printed literary labors are mostly comprised in his reports to the trustees of the hospital of which he had the superintendency for thirteen years, making in all about six hundred pages of large octavo ; a series of articles published in 1839, in the Boston *Mercantile Journal*, on the subject of an asylum for inebriates ; Hints to the Young, on a subject forced upon his attention by a large number of its unhappy victims, of both sexes, whose forlorn condition overcame his innate modesty on this subject, and induced him to publish for their warning these hints to the young ; and after leaving the hospital, his Report on the Fruits of New England was published by the Agricultural Society of Hampshire county, Massachusetts. Besides these, some minor articles of his have been printed. He wrote and delivered several lyceum lectures. His hospital reports were extensively circulated—three thousand copies being the standing order of the Legislature. While in practice, his office was for some years the great resort for young men who wished to enter the profession. It was his rule to attend to his students and hear their recitations, in the morning before sunrise. He seldom failed of being at his own house punctually at the hour of dining.

His personal appearance was commanding, and his carriage was truly majestic. His stature was six feet two and one-half inches, and without the deformity of obesity, his weight was about two hundred and sixty pounds. He was erect, and, though full in figure, his motions were quick and graceful. Although very civil and acceptable to all, he seemed born to command. Dignity and ever-enduring cheerfulness sat upon his counte-

nance, and betokened the serenity and happy state of the feelings within. •

*Resolutions on the Resignation of Dr. S. B. Woodward, offered by
Dr. J. M. Galt, May 11, 1848.*

WHEREAS, Dr. Samuel B. Woodward, at the present meeting of this
Association, has tendered his resignation as President ; therefore,

Resolved, That whilst accepting this resignation, we cannot adjourn
without declaring our high sense of the services of Dr. Woodward as
President of this body, and also our full appreciation of his ardent
and useful exertions for so many years in behalf of the unfortunate
insane.

———o———

.

The third meeting of the Association was held in New York,
on the 8th of May, 1848.   Present—

Dr. James Bates, Maine Hospital for the Insane, Augusta.

Dr. William H. Rockwell, State Asylum, Brattleboro', Vermont.

Dr. John S. Butler, Retreat for the Insane, Hartford, Connecticut.

Dr. A. Brigham, State Lunatic Asylum, Utica, New York.

Dr. C. H. Stedman, of the Boston Lunatic Hospital.

Dr. Pliny Earle, of the Bloomingdale Asylum, New York.

Dr. James Macdonald, Sanford Hall, Flushing, Long Island.

Dr. Thomas S. Kirkbride, Pennsylvania Hospital for the Insane, Philadelphia.

Dr. Joshua H. Worthington, Friends' Asylum, Frankford,
Pennsylvania.

Dr. N. D. Benedict, Philadelphia Almshouse.

Dr. John Fonerden, Maryland Hospital, Baltimore.

Dr. John M. Galt, Eastern Lunatic Asylum, at Williamsburg,
Virginia.

16

Dr. Wm. M. Awl, Lunatic Asylum, Columbus, Ohio.

Dr. John R. Allen, Lunatic Asylum, Lexington, Kentucky.

Dr. Luther V. Bell, McLean Asylum, Somerville, Massachusetts.

Dr. H. A. Buttolph, Lunatic Asylum, Trenton, New Jersey.

Dr. A. McFarland, Lunatic Asylum, Concord, New Hampshire.

Dr. N. Cutter, Private Institution, Pepperell, Massachusetts.

Dr. George H. White, Private Asylum, Hudson, New York.

Dr. M. H. Ranney, Blackwell's Island Lunatic Asylum, New York.

Dr. S. B. Woodward, having resigned his office of President, Dr. William M. Awl was elected to that position, and Dr. A. Brigham, Vice President.

Dr. Brigham read an obituary notice of Dr. Samuel White, late Vice President of the Association.

The discussson covered a great variety of subjects connected with the treatment and management of the insane, and the arrangement of hospitals for their accommodation.

The most notable feature was the adoption of a series of resolutions condemnatory of the arrangements of the lunatic asylum on Blackwell's Island, and the recommendation of many improvements in the construction and management of the institution, most of which have since been adopted.

A resolution was also adopted strongly deprecatory of the selection of medical superintendents of hospitals on political grounds, "as a dangerous departure from that sound rule which should govern every appointing power, of seeking the best men, irrespective of every other consideration."

At this meeting a resolution was passed that the trustees, managers, or official visitors of each insane asylum on this continent be invited to attend the meetings of the Association.

The Association adjourned on May 12th, to meet in Utica, New York, in 1849.

The fourth meeting of the Association was held in Utica, New York, on the 21st day of May, 1849. Present—

Dr. James Bates, Hospital for the Insane, Augusta, Maine.

Dr. L. V. Bell, McLean Asylum, Somerville, Massachusetts.

Dr. C. H. Stedman, Lunatic Hospital, Boston.

Dr. N. Cutter, Private Institution, Pepperell, Massachusetts.

Dr. I. Ray, Butler Hospital for the Insane, Providence, Rhode Island.

Dr. A. Brigham, State Lunatic Asylum, Utica, New York.

Dr. H. A. Buttolph, Lunatic Asylum, Trenton, New Jersey.

Dr. Thomas S. Kirkbride, Pennsylvania Hospital for Insane, Philadelphia

Dr. William M. Awl, Lunatic Asylum, Columbus, Ohio.

Dr. J. S. McNairy, State Hospital for Insane, Nashville, Tennessee.

Dr. C. Fremont, Beauport Asylum, Quebec, Canada East.

Dr. G. H. White, Lunatic Asylum, Hudson, New York.

Dr. W. H. Rockwell, Asylum for Insane, Brattleboro', Vermont.

Dr. C. H. Nichols, Bloomingdale Asylum, New York.

Dr. J. H. Worthington, Friends' Asylum, Frankford, Pennsylvania.

Dr. E. Jarvis, Dorchester, Massachusetts.

Dr. N. D. Benedict, Insane Department, Philadelphia Almshouse.

Resolutions expressive of the regard and esteem for Dr. James Macdonald, and regret at his death, were passed as the first act of the Association.

Dr. Bell read a paper on the disease which has been frequently called, from his description, Bell's disease.

Dr. Cutter read a history of the treatment of the insane since 1814, with special reference to his own observations in that disease.

The following resolutions are matters of interest, and were offered by Dr. Kirkbride :

Resolved, That it is the deliberate conviction of this Association that an abundance of pure air, at a proper temperature, is an essential element in the treatment of the sick, especially in hospitals, and whether for those afflicted with ordinary disease or for the insane, and that no expense that is required to effect this object thoroughly can be deemed either misplaced or injudicious.

Resolved, That the experiments recently made in various institutions in this country and elsewhere prove, to the satisfaction of the members of this Association, that the best means of supplying warmth in winter at present known to them consists in passing fresh air from the external atmosphere over pipes and plates containing steam under low pressure or hot water, the temperature of which, at the boiler, does not exceed 212° F., and placed in large air chambers in the basement or cellar of the building to be heated.

Resolved, That a complete system of forced ventilation, connected with such a mode of heating, is indispensable in every institution devoted to these purposes, and where all possible benefits are sought to be derived from its arrangements.

The following standing committees were appointed :

On the Moral Management of the Insane—Dr. Awl.

On the Medical Management of the Insane—Dr. Bates.

On the Medical Jurisprudence of Insanity—Dr. Ray.

On the Construction of Hospitals for the Insane—Dr. Kirkbride

On Restraining Apparatus—Dr. Nichols.

The Association adjourned on May 24, to meet in Boston, on the third Tuesday of June, 1850.

Dr. James Macdonald was born at White Plains, New York, July 18, 1803. Commenced the study of medicine in 1821, and received the degree of M. D. from the College of Physicians and Surgeons, of New York, on March 29, 1825, and almost immediately after was appointed resident Physician of Bloomingdale Asylum. He remained at Bloomingdale until the latter part of the year 1830, when he resigned and commenced the general practice of his profession in the city of New York. He was sent abroad in the spring of 1831, by the Governors of the New York Hospital, to visit the institutions for the insane in Europe. He returned to New York in October, 1832, and immediately took charge of the Bloomingdale Asylum, where he remained until the autumn of 1837, when he again commenced the general practice of his profession in New York city. In the following spring he was elected one of the attending physicians of the New York Hospital, which he held for four years, and then resigned. In 1839 he again visited Europe. He opened a private institution for the treatment of mental disorders, on the 1st day of June, 1841, at Murray Hill, but in the spring of 1846 removed to Sanford Hall, near Flushing, Long Island. In 1842, he was tendered the situation of Superintendent of the New York State Lunatic Asylum, which, after mature consideration, he declined.

He was attacked on the 30th of April, 1848, with severe pleuropneumonia, and died on May 2, 1849.

He materially aided in the establishment of the Asylum for the Insane on Blackwell's Island, and was appointed one of the visiting physicians in 1847.

Resolutions on the death of Dr. James Macdonald, offered by Dr. L. V. Bell, May 21, 1849:

Resolved, That as the first official act of this Association, we would give utterance to the profound sensibilities with which we have been impressed by the recent decease of our honored associate and friend, Dr. James Macdonald, of New York.

Resolved, That in view of his elevated personal character, his high intellectual attainments, his extended experience of nearly twenty-four years devoted to our department of professional labor, we deeply appreciate the breach made in the ranks of science and usefulness by his death, and in the premature close of a life of devotion to duty, at its meridian, we recognize the hand of a mysterious and inscrutable Providence, to which, however dark, we would submit, in humble faith and adoration.

Resolved, That so important an event in the history of our Association, as well as of that department of professional labor to which our lives are devoted, ought not to pass without some more enduring recognition of his life and services, and that some member be appointed to prepare and publish, and have registered in our annals, a suitable tribute to his memory, in a record of his professional life and labors.

Dr. John S. McNairy died in Nashville, Tennessee, August 18, 1849, aged thirty-seven years. He was appointed Superintendent and Physician of the State Hospital for the Insane at the age of thirty-one.

Dr. R. J. Patterson, Hospital for the Insane, Indianapolis, Indiana.

Dr. J. M. Higgins, Hospital for the Insane, Jacksonville, Illinois.

Dr. Edward Mead, Retreat for Insane, (private) Chicago, Illinois.

Dr. John R. Allen, Eastern Lunatic Asylum, Lexington, Kentucky.

Dr. John Waddell, Provincial Lunatic Asylum, St. John, New Brunswick.

Dr. James Douglass, Quebec Lunatic Asylum, Canada.

Dr. L. V. Bell was elected Vice President, in place of Dr. Brigham, deceased.

At this meeting Dr. Ray brought forward the project of a law regulating the legal relations of the insane.

A large number of valuable and interesting papers on a great variety of subjects connected with the treatment and welfare of the insane, and on the arrangement of institutions for their care, were read and discussed at this meeting.

The Association adjourned on June 22, 1850, to meet in Philadelphia on the third Monday in May, 1851.

Dr. Amariah Brigham was born at New Marlboro, Berkshire county, Massachusetts, on December 26, 1798.

He commenced practice, a youth somewhat short of his majority, in the town of Enfield, Massachusetts, where he remained two years, and then moved to Greenfield, where he continued seven years. He visited Europe, sailing on July 16, 1828, and remained abroad visiting hospitals in all the countries he visited, about a year, reaching Boston, on his return, on July 4, 1829, and in a short time resumed practice in Greenfield. He removed to Hartford, Connecticut, in April, 1831.

While in Hartford he wrote and published the following works: Influence of Mental Cultivation on Health ; Influence of Religion on the Health and Physical Welfare of Mankind ; a Treatise on Epidemic Cholera ; and an Inquiry concerning the Diseases and Functions of the Brain, the Spinal Cord, and the

Nerves. In 1837 he was elected Professor of Anatomy and Surgery in the College of Physicians and Surgeons, New York city, where he remained a year and a half. He was elected Superintendent and Physician of the Retreat for the Insane, Hartford. Connecticut, in 1840, and in the fall of 1842 to a similar appointment in connection with the State Lunatic Asylum, Utica, New York. He commenced publishing the *Journal of Insanity* in July, 1844. His health began to fail in the summer of 1847 (though he had been feeble for two years previous), and though benefited by a trip to the South, in the spring of 1848, he never fully regained it, and died September 8, 1849. In person Dr. Brigham was tall, somewhat less than six feet in height, and very slender, his weight, in health, probably not exceeding one hundred and thirty pounds. His features were well proportioned, though rather small than otherwise, eyes of a soft, dark blue, expressing more than usual the varying emotions of his mind. His hair was thin, of a brown color, and slightly, if at all, gray, at the time of his death. His gait was naturally slow, and by no means graceful, while his voice was soft, low, and quite melodious. As a whole, however, his appearance and manner indicated, to the observer, a superior and cultivated intellect, a firm will, perfect self-possession, a social disposition, and a kind and generous heart.

Resolutions on the Death of Drs. Woodward and Brigham, offered by Dr. L. V. Bell, June 21, 1850.

Resolved, That this Association has felt, beyond the power of adequate expression, the profound solemnity which has been thrown around us on the occasion of its present meeting, by the loss of two of its members, so prominent in the history of its organization, as well as in the records of the provisions for the insane in this country, and with still more sensibility, in view of the exalted personal worth, the amiable, cheerful and communicative manner, and pure, self-sacrificing lives of the deceased.

Resolved, That the deep and general regret which filled the mind of the whole philanthropic community of an entire section of country, and circles where they were best known, uttered in a thousand forms of expression, leaves us in no doubt that their virtues, merits and devotion to great public duties, have been appreciated in a degree commensurate with their just claims, and leaving neither place nor necessity for any long drawn eulogium.

Resolved. That notwithstanding the full justice which has been done to the public and private character of our distinguished friends, we still feel that the members of this Association, more intimately and fully acquainted with their peculiar traits of service and sacrifice in our specialty, ought not to be satisfied without a more particular testimonial of our feelings and opinions as to our deceased brothers; we therefore earnestly and respectfully request that Dr. Chandler would prepare, for the next meeting of the Association, a biographical sketch of the late Dr. Woodward, and that Dr. Nichols perform the same duty as regards the late Dr. Brigham.

Dr. James Bates resigned his position as Superintendent of the Maine Hospital for the Insane in 1851, and is still living at Yarmouth, Maine.

Dr. John R. Allen resigned the Superintendency of the Eastern Lunatic Asylum, of Kentucky, on October 1, 1851; moved to St. Louis, thence to Keokuk, and, after some years, to Memphis, where he is now engaged in the practice of medicine (1868).

————o————

The sixth meeting of the Association was held in Philadelphia, on May 19, 1851. Present—

Dr. Isaac Ray, Butler Hospital, Providence, Rhode Island.

Dr. N. Cutter, Pepperell, Massachusetts.

Dr. John S. Butler, Retreat for the Insane, Hartford, Connecticut.

Dr. N. D. Benedict, State Lunatic Asylum, Utica, New York.

Dr. C. H. Nichols, Bloomingdale Asylum, New York.

Dr. H. A. Buttolph, Lunatic Asylum, Trenton, New Jersey.

Dr. Thomas S. Kirkbride, Pennsylvania Hospital for the Insane, Philadelphia

Dr. J. H. Worthington, Friends' Asylum, Frankford, Pennsylvania.

Dr. William S. Haines, Insane Department, Philadelphia Almshouse.

Dr. John Curwen, Pennsylvania State Lunatic Hospital, Harrisburg.

Dr. John Fonerden, Maryland Hospital, Baltimore, Maryland.

Dr. S. Hanbury Smith, Lunatic Asylum, Columbus, Ohio.

Dr. J. W. Parker, Asylum for the Insane, Columbia, South Carolina.

Dr. R. J. Patterson, Hospital for the Insane, Indianapolis, Indiana.

Dr. J. Morrin, Lunatic Asylum, Quebec, Canada.

Dr. Pliny Earle, late of Bloomingdale Asylum, New York.

Dr. T. R. H. Smith, Lunatic Asylum, Fulton, Missouri.

Dr. J. M. Higgins, Hospital for the Insane, Jacksonville, Illinois.

Dr. W. H. Stokes, Mount Hope Institution, Baltimore, Maryland.

Dr. George Chandler, State Lunatic Hospital, Worcester, Massachusetts.

Dr. Edward Jarvis, Dorchester, Massachusetts.

Dr. M. H. Ranney, Lunatic Asylum, Blackwell's Island, New York.

Dr. Charles Evans, Consulting Physician of the Friends' Asylum, Frankford, Pennsylvania.

Lawrence Lewis, Mordecai L. Dawson, and William Biddle, managers of the Pennsylvania Hospital for the Insane, and William Bettle and John C. Allen, managers of the Friends' Asylum, Frankford; Joseph Konigmacher, trustee of the State Lunatic Hospital, Harrisburg, Pennsylvania, and Alexander Cummings, W. S. Hansell and T. Robinson, Guardians of the Philadelphia Lunatic Asylum, were invited to take seats with the members of the Association.

Dr. W. M. Awl resigned the office of President, and Dr. Luther V. Bell was elected President, and Dr. Isaac Ray, Vice President.

Resolution on Dr. Awl's resignation, offered by Dr. T. S. Kirk-bride, May 19, 1851.

Resolved, That the members of this Association, on receiving the resignation of Dr. Awl, as its presiding officer, cannot allow the occasion to pass without testifying their full appreciation of his efforts as one of the promoters of this Association, and of his varied and important services in the cause of the insane, and their regrets are increased by the knowledge that impaired health should have compelled him to cease to occupy the post of active usefulness in which he has been so long and so favorably known.

At this meeting were discussed and adopted the propositions on the construction of hospitals for the insane, which have been so eminently useful in the arrangement of hospitals to the present time :

1. Every hospital for the insane should be in the country, not within less than two miles of a large town, and easily accessible at all seasons.

2. No hospital for the insane, however limited its capacity, should have less than fifty acres of land, devoted to gardens and pleasure grounds for its patients. At least one hundred acres should be possessed by every State hospital, or other institutions for two hundred patients, to which number these propositions apply, unless otherwise mentioned.

3. Means should be provided to raise ten thousand gallons of water, daily, to reservoirs that will supply the highest parts of the building.

4. No hospital for the insane should be built without the plan having been first submitted to some physician or physicians who have had charge of a similar establishment, or are practically acquainted with all the details of their arrangements, and received his or their full approbation.

5. The highest number that can with propriety be treated in one building is two hundred and fifty, while two hundred is a preferable maximum.

6. All such buildings should be constructed of stone or brick, have slate or metalic roofs, and as far as possible, be made secure from accidents by fire.

7. Every hospital, having provision for two hundred or more patients, should have in it at least eight distinct wards for each sex—making sixteen classes in the entire establishment.

8. Each ward should have in it a parlor, a corridor, single lodging rooms for patients, an associated dormitory, communicating with a chamber for two attendants, a clothes room, a bath room, a water closet, a dining room, a dumb waiter, and a speaking tube, leading to the kitchen or other central part of the building.

9. No apartments should ever be provided for the confinement of patients, or as their lodging rooms, that are not entirely above ground.

10. No class of rooms should ever be constructed without some kind of a window in each, communicating directly with the external atmosphere.

11. No chamber for the use of a single patient should ever be less than eight by ten feet, nor should the ceiling of any story occupied by patients be less than twelve feet in height.

12. The floors of patients' apartments should always be of wood.

13. The stairways should always be of iron, stone, or other indestructible material, ample in size and number, and easy of access, to afford convenient egress in case of accident from fire.

14. A large hospital should consist of a main central building with wings.

15. The main central building should contain the offices, receiving rooms for company, and apartments, entirely private, for the superintending physician and family, in case that officer resides in the hospital building.

16. The wings should be so arranged that if rooms are placed on both sides of a corridor, the corridors should be furnished at both ends with movable glazed sashes for the free admission of both light and air.

17. The lighting should be by gas, on account of its convenience, cleanliness, safety and economy.

18. The apartments for washing clothing, &c., should be detached from the hospital building.

19. The draining should be under ground, and all the inlets to the sewers should be properly secured to prevent offensive emanations.

20. All hospitals should be warmed by passing an abundance of pure, fresh air from the external atmosphere, over pipes or plates containing steam under low pressure, or hot water, the temperature of which does not exceed 212 ° F., and placed in the basement or cellar of the building to be heated.

21. A complete system of forced ventilation, in connection with the heating, is indispensable to give purity to the air of a hospital for the insane, and no expense that is required to effect this object thoroughly, can be deemed either misplaced or injurious.

22. The boilers for generating steam for warming the building should be in a detatched structure, connected with which may be the engine for pumping water, driving the washing apparatus and other machinery.

23. All water closets should, as far as possible, be made of indestructible materials, be simple in their arrangements, and have a strong downward ventilation connected with them.

24. The floors of bath rooms, water closets and basement stories, should, as far as possible, be made of materials that will not absorb moisture.

25. The wards for the most excited class should be constructed with rooms on but one side of a corridor, not less than ten feet wide, the external windows of which should be large, and have pleasant views from them.

26. Whenever practicable, the pleasure grounds of a hospital for the insane should be surrounded by a substantial wall, so placed as not to be unpleasantly visible from the building.

These propositions were drawn up by Dr. Kirkbride.

At this meeting Dr. Kirkbride resigned the office of Secretary, which he had held since the organization of the Association, and Dr. Buttolph was elected in his place.

The following resolution, adopted at this meeting, is of interest.

Resolved, That it is the duty of the community to provide and suitably care for all classes of the insane, and that in order to secure their greatest good and highest welfare, it is indispensable that institutions for their exclusive care and treatment, having a resident medical superintendent, should be provided, and that it is improper except from extreme necessity, as a temporary arrangement, to confine insane persons in county poorhouses or other institutions, with those afflicted with or treated for other diseases or confined for misdemeanors.

The papers read and the discussions had were very interesting, varied and instructive, and the Association adjourned on May 23, 1851, to meet in New York in 1852.

Dr. William M. Awl resigned his position as superintendent of the Ohio Lunatic Asylum on July 1, 1850, and resides in Columbus, Ohio.

————o————

The seventh meeting of the Association was held in New York, commencing on May 18, 1852.   Present—

Dr. Luther V. Bell, McLean Asylum, Somerville, Massachusetts.

Dr. I. Ray, Butler Hospital, Providence, Rhode Island.

Dr. H. A. Buttolph, State Lunatic Asylum, Trenton, New Jersey.

Dr. Andrew McFarland, Insane Asylum, Concord, New Hampshire.

Dr. John S. Butler, Retreat for the Insane, Hartford, Connecticut.

Dr. Edward Jarvis, Dorchester, Massachusetts.

Dr. N. Cutter, Pepperell, Massachusetts.

Dr. C. A. Walker, Lunatic Hospital, Boston, Massachusetts.

Dr. C. H. Nichols, Bloomingdale Asylum, New York.

Dr. N. D. Benedict, State Lunatic Asylum, Utica, New York.

Dr. M. H. Ranney, Lunatic Asylum, Blackwell's Island, New York.

Dr. Thomas S. Kirkbride, Pennsylvania Hospital for the Insane, Philadelphia.

Dr. J. H. Worthington, Friends' Asylum, Frankford, Pennsylvania.

Dr. John Curwen, Pennsylvania State Lunatic Hospitál, Har-. risburg.

Dr. Francis T. Stribling, Western Lunatic Asylum, Staunton, Virginia.

Dr. S. Hanbury Smith, Lunatic Asylum, Columbus, Ohio.

Dr. Thomas F. Green, State Lunatic Asylum, Milledgeville, Georgia.

Dr. Francis Bullock, Kings County Lunatic Asylum, Flatbush, New York.

Dr. A. Lopez, Hospital for the Insane, Alabama.

Drs. Henry W. Buell and B. Ogden, Sanford Hall, Flushing, New York.

Dr. J. M. Higgins, Hospital for the Insane, Jacksonville, Illinois.

Dr. George Chandler, State Lunatic Hospital, Worcester, Massachusetts.

Dr. William S. Haines, Insane Department, Philadelphia Almshouse.

Dr. C. Fremont, Lunatic Asylum, Quebec, Canada.

Dr. R. J. Patterson, Hospital for the Insane, Indianapolis, Indiana.

Visitors—E. A. Wetmore, Esq., Treasurer State Lunatic Asylum, Utica.

Stacy B. Collins, Board of Governors of New York Hospital.

A. Munson, Esq., President of Board of Managers of State Lunatic Asylum, Utica, New York.

The discussions of the various papers presented were interesting and instructive, and afier a very pleasant meeting, the Association adjourned on May 22, 1852, to meet in Baltimore, Maryland.

Dr. J. M. Higgins was removed from his position by a new board of Trustees of the Illinois Hospital, on June 6, 1853, and is still living, engaged in the practice of medicine in Griggsville, Illinois.

——o——

The eighth meeting was held in Baltimore, Maryland, commencing on May 10, 1853. Present—

Dr. Thomas S. Kirkbride, Pennsylvania Hospital for the Insane, Philadelphia.

Dr..F. T. Stribling, Western Lunatic Asylum, Staunton, Virginia.

Dr. N. D. Benedict, State Lunatic Asylum, Utica, New York.

Dr. H. A. Buttolph, State Lunatic Asylum, Trenton, New Jersey.

Dr. D. Tilden Brown, Bloomingdale Asylum, New York.

Dr. John Curwen, Pennsylvania State Lunatic Hospital, Harrisburg.

Dr. J. H. Worthington, Friends' Asylum, Frankford, Pennsylvania.

Dr. John E. Tyler, Asylum for the Insane, Concord, New Hampshire.

Dr. R. J. Patterson, Hospital for the Insane, Indianapolis, Indiana.

Dr. Elijah Kendrick, Lunatic Asylum, Columbus, Ohio.

Dr. Clement A. Walker, Lunatic Hospital, Boston, Massachusetts.

Dr. John Fonerden, Maryland Hospital, Baltimore, Maryland.

Dr. William H. Stokes, Mount Hope Institution, Baltimore, Maryland.

Dr. Francis Bullock, Kings county Lunatic Asylum, Flatbush, New York.

Dr. Luther V. Bell, McLean Asylum, Somerville, Massachusetts.

Dr. I. Ray, Butler Hospital, Providence, Rhode Island.

Dr. Edward Jarvis, Dorchester, Massachusetts.

Dr. J. D. Stewart, Insane Department, Philadelphia Almshouse.

Dr. T. R. H. Smith, State Lunatic Asylum, Fulton, Missouri.

Dr. C. H. Nichols, Government Hospital for the Insane, Washington, District of Columbia.

Many interesting papers were read at this meeting, but the most important action was the adoption of the proposition on organization of Hospitals for the Insane :

1. The general controlling power should be vested in a board of trustees or managers; if of a State institution, selected in such a manner as will be likely most effectually to protect it from all influences connected with political measures or political changes ; if of a private corporation, by those properly authorized to vote.

2. The board of trustees should not exceed twelve in number, and be composed of individuals possessing the public confidence, distinguished for liberality, intelligence, and active benevolence, above all political influence, and able and willing faithfully to attend to the duties of their station. Their tenure of office should be so arranged, that where changes are deemed desirable, the terms of not more than one-third of the whole number should expire in one year.

3. The board of trustees should appoint the physician, and on his nomination, and not otherwise, the assistant physician, steward and matron. They should, as a board or by committee, visit and examine every part of the institution, at frequent stated intervrls, not less than semi-monthly, and at such other times as they may deem expedient, and exercise so careful a supervision over the expenditures and general operations of the hospitals, as

to give the community a proper degree of confidence in the correctness of its management.

4. The physician should be the superintendent and chief executive officer of the establishment. Besides being a well educated physician, he should possess the mental, physical and social qualities to fit him for the post. He should serve during good behavior, reside on or very near the premises, and his compensation should be so liberal as to enable him to devote his whole time and energies to the welfare of the hospital. He should nominate to the board suitable persons to act as assistant physician, steward and matron; he should have the entire control of the medical, moral and dietetic treatment of the patients, the unreserved power of appointment and discharge of all persons engaged in their care, and should exercise a general supervision and direction of every department of the institution.

5. The assistant physician, or assistant physicians, where more than one are required, should be graduates of medicine, of such character and qualifications as to be able to represent and perform the ordinary duties of the physician during his absence.

6. The steward, under the direction of the superintending physician, and by his order should make all purchases for the institution, keep the accounts, make engagements with, and pay and discharge those employed about the establishment; have a supervision of the farm, garden and grounds, and perform such other duties as may be assigned him.

7. The matron, under the direction of the superintendent, should have a general supervision of the domestic arrangements of the house, and, under the same direction, do what she can to promote the comfort and restoration of the patients.

8. In institutions containing more than two hundred patients, a second assistant physician and apothecary should be employed, to the latter of whom other duties, in the male wards, may be conveniently assigned.

9. If a chaplain is deemed desirable as a permanent officer, he should be selected by the superintendent, and like all others

engaged in the care of the patients, should be entirely under his direction.

10. In every hospital for the insane, there should be one supervisor for each sex. exercising a general oversight of all the attendants and patients, and forming a medium of communication between them and the officers.

11. In no institution should the number of persons in immediate attendance on the patients be in a lower ratio than one attendant for every ten patients, and a much larger proportion of attendants will commonly be desirable.

12. The fullest authority should be given to the superintendent to take every precaution that can guard against fire or accident within an institution, and to secure this an efficient night watch should always be provided.

13. The situation and circumstances of different institutions may require a considerable number of persons to be employed in various other positions, but in every hospital, at least all those that have been referred to are deemed not only desirable, but absolutely necessary, to give all the advantages that may be hoped for from a liberal and enlightened treatment of the insane.

14. All persons employed in the care of the insane should be active, vigilant, cheerful and in good health. They should be of a kind and benevolent disposition, be educated, and in all respects trustworthy, and their compensation should be sufficiently liberal to secure the services of individuals of this description.

These propositions were prepared by Dr. Kirkbride.

During this meeting the Association visited Washington, to examine the site of the new hospital for the army and navy and District of Columbia.

The Association adjourned on May 13, 1853, to meet in Washington, District of Columbia.

Dr. N. D. Benedict was obliged to resign his position as Superintendent of the New York State Lunatic Asylum, to which he had been appointed in the fall of 1849. He had previously held the office of chief physician in the Philadelphia Almshouse,

34

and had commenced and put in operation great charges in the
insane department of that institution. Since leaving Utica, he
has resided, on account of his tendency to disorder of the lungs,
in Florida.

Dr. J. D. Stewart died on April 12, 1854, in the fortieth year
of his age.

Dr. Elijah Kendrick resigned his position as Superintendent
of Central Ohio Lunatic Asylum on July 1, 1854, and for several
years past has had a private institution near New Brighton,
Beaver county, Pennsylvania.

Dr. Francis Bullock was born at Centreville, Allegheny county,
New York, June 22, 1828 ; received his degree of M. D. from the
College of Physicians and Surgeons of New York, in October,
1849 ; was appointed resident physician of Kings County Lunatic Asylum, at Flatbush, Long Island, January, 1850, which place
he held at the time of his death, in July, 1853.

——o——

The ninth annual meeting was held in the city of Washington,
District of Columbia, in the rooms of the Smithsonian Institute,
commencing on May 9, 1854. The following members were
present :

Dr. Luther V. Bell, McLean Asylum, Somerville, Massachusetts.

Dr. Isaac Ray, of the Butler Hospital, Providence, Rhode
Island.

Dr. F. T. Stribling, of the Western Lunatic Asylum, Staunton, Virginia.

Dr. Thomas S. Kirkbride, Pennsylvania Hospital for the Insane, Philadelphia, Pennsylvania.

Dr. T. R. H. Smith, State Lunatic Asylum, Fulton, Missouri.

Dr. James S. Athon, Hospital for the Insane, Indianapolis, Indiana.

Dr. John Waddell, Provincial Lunatic Asylum, St. John, New Brunswick.

Dr. John Curwen, Pennsylvania State Lunatic Hospital, Harrisburg.

Dr. Edward C. Fisher, Insane Asylum, Raleigh, North Carolina.

Dr. W. A. Cheatham, Hospital for the Insane, Nashville, Tennessee.

Dr. John E. Tyler, Asylum for the Insane, Concord, New Hampshire.

Dr. William H. Stokes, Mount Hope Institution, Baltimore, Maryland.

Dr. J. H. Worthington, Friends' Asylum, Frankford, Philadelphia, Pennsylvania.

Dr. C. A. Walker, Boston Lunatic Hospital, Boston, Massachusetts.

Dr. D. T. Brown, Bloomingdale Asylum, New York.

Dr. John Fonerden, Maryland Hospital, Baltimore, Maryland.

Dr. Edward Jarvis, Dorchester, Massachusetts.

Dr. Joseph Morrin, Lunatic Asylum, Quebec, Canada East.

Dr. T. M. Ingraham, Kings County Lunatic Asylum, Flatbush, New York.

Dr. M. H. Ranney, New York City Lunatic Asylum, Blackwell's Island.

Dr. C. H. Nichols, Government Hospital for the Insane, Washington, District of Columbia.

Dr. William M. Awl, late of Lunatic Asylum, Columbus, Ohio.

Dr. Buttolph resigned his office of Secretary, and Dr. Nichols was chosen to that office.

The Association paid their respects to the President of the United States, visited Mount Vernon, and also the principal public buildings in Washington, and after a pleasant and profitable meeting, adjourned on May 12, 1854, to meet in the city of Boston.

Dr. T. M. Ingraham left the institution of which he was for a short time physician (Kings County Lunatic Asylum), to engage in private practice in Flatbush, and is now (1868) in practice at Flatlands. New York.

————o————

The tenth annual meeting of the Association was held in Boston, Massachusetts, commencing on May 22, 1855. The use of the Senate Chamber having been tendered to the Association by that body, the Association was organized there. The following members were present :

Dr. Luther V. Bell, McLean Asylum, Somerville, Massachusetts.

Dr. Isaac Ray, Butler Hospital, Providence, Rhode Island.

Dr. T. S. Kirkbride, Pennsylvania Hospital for the Insane, Philadelphia

Dr. C. H. Nichols, Government Hospital for the Insane, Washington, District of Columbia.

Dr. John S. Butler, Retreat for the Insane, Hartford, Connecticut.

Dr. John Curwen, Pennsylvania State Lunatic Hospital, Harrisburg, Pennsylvania.

Dr. H. A. Buttolph, Lunatic Asylum, Trenton, New Jersey.

Dr. J..H. Worthington, Friends' Asylum, Philadelphia, Pennsylvania.

Dr. W. H. Rockwell, Asylum for the Insane, Brattleboro', Vermont.

Dr. James S. Athon, Hospital for the Insane, Indianapolis, Indiana.

Dr. T. R. H. Smith, State Lunatic Asylum, Fulton, Missouri.

Dr. Edward Jarvis, Dorchester, Massachusetts.

Dr. D. T. Brown, Bloomingdale Asylum, New York.

Dr. N. Cutter, Pepperell, Massachusetts.

Dr. C. H. Stedman, Boston, Massachusetts.

Dr. H. M. Harlow, Insane Hospital, Augusta, Maine.

Dr. E. S. Blanchard, Kings County Lunatic Asylum, Flatbush, New York.

Dr. John E. Tyler, Asylum for the Insane, Concord, New Hampshire.

Dr. G. C. S. Choate, State Lunatic Hospital, Taunton, Massachusetts.

Dr. John P. Gray, State Lunatic Asylum, Utica, New York.

Dr. Edward C. Fisher, Asylum for the Insane, Raleigh, North Carolina.

Dr. C. A. Walker, Boston Lunatic Hospital, South Boston, Massachusetts.

Dr. Joseph Workman, Provincial Lunatic Asylum, Toronto, Canada West.

Dr. George Chandler, State Lunatic Hospital, Worcester, Massachusetts.

Dr. M. H. Ranney, New York City Lunatic Asylum, Blackwell's Island.

Dr. Joshua Clements, Southern Ohio Lunatic Asylum, Dayton.

Also, Dr. George Dock, one of the Trustees of the Pennsylvania State Lunatic Hospital, Harrisburg, who remained in attendance till the final adjournment.

Dr. Bell resigned the office of President, and Dr. Ray was elected President, and Dr. Kirkbride Vice President. Dr. Butler was elected Treasurer in place of Dr. Kirkbride.

Resolution on Dr. L. V. Bell's resignation, offered by Dr. T. S. Kirkbride, May 22, 1855.

Resolved, That the Association has accepted with regret the resignation of Dr. Bell, and that the thanks of the Association be tendered to him for the able manner in which he has performed the duties of his station.

Dr. Ray read his paper on the Insanity of George the Third.

During this session an invitation was presented by the Mayor of the city, to the Association, to occupy the room of the Common Council, in the City Hall, for the sessions of the Association, which was accepted, and the subsequent meetings were held in that room.

More than the usual number of papers on interesting subjects were read and discussed, and the Association was the recipient of unusual attention from the city authorities of Boston, to whom they were indebted for a steamboat excursion down the bay, and to the various public buildings and objects of interest ; and also from the officers of different public institutions in Boston and its vicinity.

Dr. Jarvis presented to the Association, in a condensed form, the results of the investigation of a commission to ascertain the number of the insane in Massachusetts.

The Association adjourned on May 25, to meet in Cincinnati, Ohio, on the third Tuesday of May, 1856.

Dr. Blanchard remained in connection with the Kings County Lunatic Asylum about one year, then engaged in practice in New York, and subsequently moved to Vermont.

Dr. Geo. Chandler resigned his office in the fall of 1855, and has since been living in Worcester, Massachusetts.

Dr. C. H. Stedman, after leaving the Boston Lunatic Hospital, settled in Boston, and engaged in general practice, filling various public positions with credit and honor, and died on June 8, 1866.

Dr. Luther V. Bell was born in Chester, New Hampshire, December 30, 1806 , son of Hon. Samuel Bell, successively Chief Justice, Governor, and United States Senator from New Hampshire.

He entered Bowdoin College at twelve years of age, and graduated in 1823, He received his medical degree from Dartmouth College in 1826, and subsequently pursued his medical studies in Europe. He commenced and pursued the practice of medicine and surgery in the towns of Brunswick and Derry, New Hampshire, with success in both departments, and interested himself largely in sanitary and philanthropic measures, tending to the elevation of his profession and the general welfare of the people. In 1834 he was awarded the Boylston Prize Medal for a dissertation on the dietetric regimen best fitted for the inhabi-

tants of New England. In 1835 he presented an essay on the
External Exploration of Diseases, which forms the first third of
the ninth volume of the Library of Practical Medicine. He
subsequently put forth a small volume, entitled, An attempt to
investigate some obscure and undecided doctrines in relation to
small pox and varioliform diseases. He labored earnestly in the
establishment of the New Hampshire Asylum for the Insane, was
elected to the General Court with the special object of urging
forward this measure, and made a very able report on the number and condition of the insane of that State, and the means of
providing for them. While attending a second session of the
Legislature and pressing this object, he received very unexpectedly
the intelligence of his having been appointed physician and Superintendent of the McLean Asylum for the Insane. He was appointed during the latter part of 1836, and entered upon his
official duties at the beginning of the next year.

He was an early and earnest advocate for the introduction of
steam and hot water, and mechanical power, as the proper and
only suitable mode of warming and ventilating hospitals, and the
McLean Asylum, over which he presided, was the first institution
in which a circulation of hot water was successfully employed for
warming a large inflowing current of air.

In 1845, on the solicitation of the Trustees of the Butler Hospital for the Insane, at Providence, Rhode Island, then in con-
templation, the Trustees of the Massachusetts General Hospital,
of which the McLean Asylum is a branch, gave Dr. Bell leave of
absence to visit Europe, that he might, after a comparison of the
institutions of the old world, be enabled to devise a plan of hospital embodying all that was excellent and desirable then known
to the profession. After his return he presented the plan of that
establishment which has so fully met the highest hopes of his
friends.

He was for two years President of the Massachusetts Medical
Society, and his inaugural address was on Ventilation.

He subsequently published a small volume, entitled, The practical method of ventilating buildings, with an appendix on heating by steam and hot water.

He was one of the original members of this Association, and its President from 1850 to 1855.

He held the post of Executive Councillor in the administration of Governor Briggs, in 1850, and was a member of the Committee of Pardons, to which was referred two cases famous in the annals of crime in Massachusetts. That of Daniel Pearson, convicted of the murder of his wife and infant twin children, and that of Prof. J. W. Webster, for the murder of Dr. George Parkman. He was the candidate of the Whig party in the Seventh Congressional District of Massachusetts, in 1853, but though receiving a plurality of votes in the first trial, was beaten on the second by the union of the two opposing parties on the same candidate.

He was also a delegate to the Convention for revising the State Constitution.

He resigned his position as Superintendent of the McLean Asylum in the fall of 1856, the state of his health urging this step. In addition to impaired health from pulmonary disease, he had lost children one after another, at the most touching epochs of parental attachment, and under the highest hopes. The death of his estimable wife filled the measure of his domestic sorrow. From the McLean Asylum he removed to his private residence, in Monument Square, Charlestown.

Here his life was not a retirement, as he was constantly consulted in cases of insanity and other cerebral and nervous affections, and on questions of medico-legal character.

At the breaking out of the rebellion, he was among the first to offer his services to the Government. He went as Surgeon os the 11th Regiment of Massachusetts Volunteers, but was soon promoted to the position of Brigade Surgeon to Gen. Hooker's Division, on the Lower Potomac.

He died in camp quite suddenly from endocarditis, on February 11, 1862.

He was known to the older members of this Association as the able alienist physician, his great skill in the detection of disordered mental manifestations, by his elaborate description of that form of acute mania, so often described as Bell's disease, by his genial qualities and his earnest persevering efforts to advance the specialty to the highest rank.

————o————

The eleventh annual meeting of the Association was held at the Spencer House, in the City of Cincinnati, commencing at 10 o'clock A. M., of May 19, 1856. The following members were present :

Dr. John Fonerden, Maryland Hospital, Baltimore, Maryland.

Dr. Joseph Workman, Provincial Lunatic Asylum, Toronto, Canada.

Dr. Andrew McFarland, Hospital for the Insane, Jacksonville, Illinois.

Dr. James S. Athon, Hospital for the Insane, Indianapolis, Indiana.

Dr. J. J. McIlhenney, Superintendent elect of the Southern Ohio Lunatic Asylum, Dayton, Ohio.

Dr. Joshua Clements, Southern Ohio Lunatic Asylum, Dayton, Ohio.

Dr. D. Tilden Brown, Bloomingdale Asylum, Manhattanville, New York.

Dr. J. J. Quinn, Hamilton County Lunatic Asylum, Cincinnati, Ohio.

Dr. O. M. Langdon, Superintendent elect of Hamilton County Lunatic Asylum, Cincinnati, Ohio.

Dr. John Curwen, Pennsylvania State Lunatic Hospital, Harrisburg, Pennsylvania.

Dr. John P. Gray, New York State Lunatic Asylum, Utica, New York.

Dr. William A. Cheatham, Tennessee Hospital for the Insane, Nashville, Tennessee.

Dr. Edward Mead, Retreat for the Insane, Cincinnati, Ohio.

Dr. T. R. H. Smith, State Lunatic Asylum, Fulton, Missouri.

Dr. C. H. Nichols, Government Hospital for the Insane, Washington, District of Columbia.

Dr. E. H. VanDeusen, Michigan Asylum for the Insane, Kalamazoo.

Dr. Isaac Ray, Butler Hospital, Providence, Rhode Island.

Dr. R. Hills, Central Ohio Lunatic Asylum, Columbus, Ohio.

Dr. G. E. Ells, late of Central Ohio Lunatic Asylum, Columbus, Ohio.

Dr. J. H. Worthington, Friends' Asylum, Frankford, Pennsylvania.

Dr. R. C. Hopkins, Superintendent elect of Newburg (O.) Lunatic Asylum, Newburg, Ohio.

Dr. Joseph A. Reed, Western Pennsylvania Hospital for the Insane, Pittsburg, Pennsylvania.

Dr. M. H. Ranney, New York City Lunatic Asylum, New York.

Dr. W. S. Chipley, Eastern Lunatic Asylum, Lexington, Kentucky.

Dr. John E. Tyler, Lunatic Asylum, Concord, New Hampshire.

Dr. R. B. Baisely, Kings County Lunatic Asylum, Flatbush, New York.

Dr. S. Hanbury Smith, formerly of the Columbus Lunatic Asylum, Columbus, Ohio.

Dr. G. C. S. Choate, State Lunatic Hospital, Taunton, Massachusetts.

In the absence of the President, Dr. Fonerden was appointed President *pro tem.*

The number of papers read on various subjects connected with the specialty was large, and the discussions earnest and interesting. The plans of the new Institution for the Insane of Hamilton county, near Cincinnati, were laid before the Association for examination, and such suggestions as the members might think proper to make. The plans were referred to a special committee, who reported various alterations and improvements in the plans, and also recommended that the Architect be authorized by the Commissioners to visit the different institutions in actual operation.

The Association were the recipients of numerous courtesies and attentions from the trustees of the different institutions in and around Cincinnati, and also from several gentlemen having large vineyards which they had full opportunities of examining.

The Association adjourned on May 22, 1856, to meet in New York.

Dr. J. J. Quinn is still living and practicing medicine in Cincinnati.

Dr. Geo. E. Ells died in the fall of 1867, of Bright's disease.

Dr. R. B. Baisely was the first physician who had charge of the present Asylum at Flatbush, into which the patients were moved in October, 1855, and remained about eighteen months, and then went into practice in Rockaway, Long Island, where he now lives.

————o————

The twelfth annual meeting was held at the Metropolitan Hotel, in the city of New York, commencing at 10 A. M., on May 19, 1857. The following members were present :

Dr. John E. Tyler, Asylum for the Insane, Concord, New Hampshire.

Dr. J. P. Bancroft, Superintendent elect Asylum for the Insane, Concord, New Hampshire.

Dr. W. H. Rockwell, Asylum for Insane, Brattleboro', Vermont.

Dr. Chauncey Booth, McLean Asylum, Somerville, Massachusetts.

Dr. Merrick Bemis, State Lunatic Hospital, Worcester, Massachusetts.

Dr. G. C. S. Choate, State Lunatic Hospital, Taunton, Massachusetts.

Dr. Clement A. Walker, Boston Lunatic Hospital, South Boston.

Dr. Edward Jarvis, Dorchester, Massachusetts.

Dr. N. Cutter, Pepperell, Massachusetts.

Dr. I. Ray, Butler Hospital for the Insane, Providence, Rhode Island.

Dr. John S. Butler, Retreat for the Insane, Hartford, Connecticut.

Dr. D. Tilden Brown, Bloomingdale Asylum, New York.

Dr. J. W. Barstow, Sanford Hall, Flushing, Long Island.

Dr. Pliny Earle.

Dr. H. W. Buell.

Dr. M. H. Ranney, New York City Lunatic Asylum, New York.

Dr. John V. Lansing, Kings county Lunatic Asylum, Flatbush, Long Island.

Dr. Benjamin Ogden, Sanford Hall, Flushing, Long Island.

Dr. H. A. Buttolph, State Lunatic Asylum, Trenton, New Jersey.

Dr. Thomas S. Kirkbride, Pennsylvania Hospital for the Insane, Philadelphia.

Dr. Joshua H. Worthington, Friends' Asylum, Frankford, Pennsylvania.

Dr. John Curwen, Pennsylvania State Lunatic Hospital, Harrisburg.

Dr. Joseph A. Reed, Western Pennsylvania Hospital for the Insane, Pittsburg, Pennsylvania.

Dr. John Fonerden, Maryland Hospital, Baltimore.

Dr. William H. Stokes, Mount Hope Institution, Baltimore, Maryland.

Dr. C. H. Nichols, Government Hospital for the Insane, Washington, District of Columbia.

Dr. Edward C. Fisher, Asylum for the Insane, Raleigh, North Carolina,

Dr. R. C. Hopkins, Northern Ohio Lunatic Asylum, Newburg, Ohio.

Dr. J. J. McIlhenny, Southern Ohio Lunatic Asylum, Dayton, Ohio.

Dr. William Mount, Hamilton County Lunatic Asylum, near Cincinnati, Ohio.

Dr. James S. Athon, Hospital for the Insane, Indianapolis, Indiana.

Dr. Andrew McFarland, Hospital for the Insane, Jacksonville, Illinois.

Dr. E. H. VanDeusen, Asylum for Insane, Kalamazoo, Michigan.

Dr. James Douglas, Lunatic Asylum, Quebec, Canada East.

Dr. James A. DeWolf, Provincial Asylum for the Insane, Halifax, Novia Scotia.

An interesting incident of the meeting was the presentation, for examination by the members, of the photographs of four generations of the Tuke family, commencing with William Tuke, who took so prominent a part in the amelioration of the condition of the insane in England, in 1788.

The plan of the department for males of the Pennsylvania Hospital for the Insane, then in course of erection ; and also the plan for a new institution for the criminal insane, at Auburn, New York, were laid before the Association for examination.

After the reading and discussion of many interesting papers, and the visiting of the different institutions in and around the city of New York, and also the reception of many courtesies from various gentlemen, the Association adjourned on May 22, 1857, to meet in Quebec, in June, 1858.

Dr. Chauncey Booth died on January 12, 1858, aged 41 years.

Dr. Booth had suffered under marked pulmonary disease ever since the winter of 1850–51. Cavities in one lung were distinctly diagnosed as far back as that date, and the evidence of slow but continuous progress were manifest until the scene closed. If

there were ever an unequivocal example of will-power, in sus-
pending and retarding the certain march of phthisis, it was in
this case. Looking his symptoms daily in the face, he seemed
to feel that he had an enemy to be met, and that every foot of
ground was to be contested with him. As brave as any hero who
ever faced the cannon's mouth, he never allowed his stern and
unrelenting foe to gain upon him by intimidation. He kept
coolly at work, subverting the approach of the enemy by every
strategic means which science and experience furnished to his aid,
but no panic, no disheartening yielding, ever lost him an inch in
the contest. And as if to determine a victory in favor of the
unintimidated contestant, phthisis did *not* win its usually easy and
certain triumph. Two months before Dr. Booth's disease, when
the consumptive symptons had scarcely a more prominent place
than they had had for six or eight years, Bright's disease set in
with its distinct features. The noble victim recognized the fatal
weight of this unexpected ally, and calmly yielded to the over-
whelming forces of the combined enemy.

The immediate approach of death was met in the same spirit
which had marked the entire onward march of the enemy. There
was neither bravado, nor boast, nor affected indifference. He
set his house in order as deliberately as one arranges for a dis-
tant journey, and when the last moments were approaching, he
desired that his only child, a boy of some seven or eight sum-
mers, weeping at the scene, should be removed so as to escape
the lasting impression of the physical effects of the struggle, "*in
articulo mortis.*"

He had been assistant physician at Brattleboro', Augusta, and
at Somerville. He did not leave much for the literature of our
specialty. Beginning our work at twenty years of age, he labored without intermission with us to the close, and he never until the last two years, when every moment was crowded with
duties, would have consented to put himself forward as an in-
structor of others. And this modesty was perfectly sincere. In
1847 he drew up, and that only by request, an account of an
epidemic dysentery of some eighty cases, at the Asylum, which
commanded the highest encomiums of the late Dr. Fisher, the

best pathologist of his time and place. Dr. Booth's only hospital report will stand as a bright memorial of what the man *was*, while, as the trustees in their report indicate, his papers in their files demonstrate what he would have been as chief of a great hospital for the insane.

A striking feature of his personal character was his eminent, social, genial wit, an instinctive power of seizing and grouping together the most unexpected and incongruous images, all most telling and illustrative of the subject matter in point, yet unlike the almost inseperable incident of the ordinary possession of this dangerous gift, never leaving behind one sting, or a single allusion which any party could repeat.

He went on through life, not merely "without an enemy," giving the idea in its stale and well-worn phrase, but absolutely without a suspicion of what an enemy might be.

A remarkable feature of Dr. Booth's character was, that while he had never been "in the world," he had as complete and sagacious an idea of its entire system, as if he had plunged into the perplexities of trade, the struggles of ambition, and the debasement of the passions. He passed from the pure circle of the family of a Connecticut clergyman, the father as marked for a holy simplicity, as the mother was for the traits which characterized the son, into the wards of a great lunatic hospital, thence to another, thence to a third, and thence—to his reward. No man of the age of forty, in this community, can be found on search, who ever passed so few days away from the immediate fields of his daily duty.

Like all other men devoted to one absorbing pursuit, he had his own pet pleasures, his peculiar side avocations, to which he loved to steal after every call of duty was over, and in the stillness of the household fireside. Yet few men of that great company of those who knew him in the same pursuit as themselves, could probably ever have conjectured wherein that specific taste would have shown itself. It was in the study of the ecclesiastical history of New England.

The strong point of Dr. Booth's professional character was an

absolute identification with the insane. If not born within hospital walls, he had passed his whole actual life within them, and
never seemed to dream of being anywhere else. No man seemed
so perfectly to enter into the insane nature of those around him.

Buried with him in the quiet shades of the Cemetery of
Mount Auburn was no common measure of that mighty talent
of dealing with the insane mind, which, as was well observed by
one of the great masters of our art, "can be acquired, but never
can be communicated. It must die with its possessor."

*Resolutions on the Death of Dr. Chauncey Booth, offered by Dr. J.
E. Tyler, June 8, 1858 :*

Resolved, That the death of Dr. Chauncey Booth is felt to be an
irreparable loss to this Association, and that we offer to his family
our sincere sympathy and condolence in this our mutual bereavement.

Dr. Nehemiah Cutter died at Pepperell, Massachusetts, on
March 15, 1859. Dr Cutter was a native of New Hampshire,
and a graduate of Dartmouth College. His name had for more
than forty years been known in connection with the Private Asylum for Nervous Invalids, established by him at Pepperell, and
he was also one of the original members of the Association. An
incident which occurred near the close of Dr. Cutter's life, best
illustrates the character and ability of the man. In a single hour
the devouring elements laid in ashes the accumulation of a laborious life. In every sense of the word his occupation seemed
to be gone. To rebuild for the same purpose would have been
out of the question. Nothing daunted, however, he assumed
immediately the long laid aside duties of common professional
life, and won as a practicing physician, when close upon three
score and ten, the fresh confidence of the community in which
he lived and died.

———o———

The thirteenth meeting of the Association was held at Russell's hotel, in the city of Quebec, Canada East, commencing at 10 o'clock A. M., of June 8, 1858. The following members were present :

Dr. Isaac Ray, of the Butler Hospital for the Insane, Providence, Rhode Island.

Dr. Thomas F. Green, State Lunatic Asylum, Milledgeville, Georgia.

Dr. William B. Williamson, State Lunatic Asylum, Jackson, Mississippi.

Dr. James S. Athon, Hospital for the Insane, Indianapolis, Indiana.

Dr. Joseph Workman, Provincial Lunatic Asylum, Toronto, Canada.

Dr. W. S. Chipley, Eastern Lunatic Asylum, Lexington, Kentucky.

Dr. James Douglas, Joseph Morrin and C. Fremont, of the Quebec Lunatic Asylum.

Dr. T. R. H. Smith, State Lunatic Asylum, Fulton, Missouri.

Dr. J. V. Lansing, Kings County Lunatic Asylum, Flatbush, New York.

Dr. G. C. S. Choate, State Lunatic Hospital, Taunton, Massachusetts.

Dr. J. J. McIlhenney, Southern Ohio Lunatic Asylum, Dayton, Ohio.

Dr. R. Hills, Central Ohio Lunatic Asylum, Columbus, Ohio.

Dr. J. H. Worthington, Friends' Asylum, Frankford, Philadelphia, Pennsylvania.

Dr. William Mount, Hamilton County Lunatic Asylum, Cincinnati, Ohio.

Dr. E. H, Van Deusen, Asylum for the Insane, Kalamazoo, Michigan.

Dr. Edward A. Smith, Pennsylvania Hospital for the Insane, Philadelphia, Pennsylvania.

Dr. John Curwen, Pennsylvania State Lunatic Hospital, Harrisburg, Pennsylvania.

Dr. John E. Tyler, McLean Asylum, Somerville, Massachusetts.

Dr. W. H. Rockwell, Hospital for the Insane, Brattleboro', Vermont.

Dr. M. Bemis, State Lunatic Hospital, Worcester, Massachusetts.

Dr. H. M. Harlow, Hospital for the Insane, Augusta, Maine.

Dr. Andrew McFarland, Hospital for the Insane, Jacksonville, Illinois.

Dr. Edward Jarvis, Dorchester, Massachusetts.

Dr. M. H. Ranney, New York City Lunatic Asylum.

Dr. Samuel Grimes and Henry Brady, Esq., Commissioners of the Indiana Hospital for the Insane, and Governors W. F. Pickney and B. F. Pickney, of the New York City Lunatic Asylum, were invited to attend the meeting of the Association.

Dr. Nichols resigned the Secretaryship of the Association, and Dr. Curwen was elected in his place.

Many papers of great interest and value were read and discussed ; and under the courteous guidance of Drs. Douglas, Morrin and Fremont, the members had the opportunity of visiting all the public institutions of the city of Quebec, and also all objects of interest in the neighborhood.

The Association adjourned on Thursday, June 10, 1858, to meet in Lexington, Kentucky.

Dr. William B. Williamson was removed, during the year 1858, from the Superintendency of the Mississippi Hospital for the Insane.

Dr. J. V. Lansing resigned in 1853, and has since been in practice in Albany, New York.

Dr. R. C. Hopkins was elected Assistant Physician of the Central Ohio Lunatic Asylum in the spring of 1844, and remained there four years. From that time, till the spring of 1856, he was in private practice in Cleveland, Ohio. In April, 1856, he was chosen Superintendent of the Northern Ohio Lunatic Asylum, at Newburg, where he remained till December, 1858, when he again engaged in general practice. In 1862 he entered the service of the United States Sanitary Commission as Medical Inspector, and it was in the labor of establishing a hospital for soldiers at Memphis, Tennessee, that he contracted the disease—typhoid pneumonia—of which he died.

——o——

The fourteenth annual meeting was held at the Phœnix Hotel in the city of Lexington, Kentucky, commencing at 10 o'clock A. M. of May 17, 1859. The following members were present :

Dr. James S. Athon, Hospital for the Insane, Indianapolis, Indiana.

Dr. J. D. Barkdull, Insane Asylum, Jackson, Louisiana.

Dr. G. C. S. Choate, State Lunatic Hospital, Taunton, Massachusetts.

Dr. W. S. Chipley, Eastern Lunatic Asylum, Lexington, Kentucky.

Dr. W. A. Cheatham, State Lunatic Hospital, Nashville, Tennessee.

Dr. John P. Gray, State Lunatic Asylum, Utica, New York.

Dr. R. Hills, Central Ohio Lunatic Asylum, Columbus, Ohio.

Dr. O, C. Kendrick, Northern Ohio Lunatic Asylum, Newburg, Ohio.

Dr. Robert Kells, State Lunatic Asylum, Jackson, Mississippi.

Dr. Andrew McFarland, Hospital for the Insane, Jacksonville, Illinois.

Dr. J. J. McIlhenney, Southern Ohio Lunatic Asylum, Dayton, Ohio.

Dr. William Mount, Hamilton County Lunatic Asylum, Cincinnati, Ohio.

Dr. F. G. Montgomery, Western Lunatic Asylum, Hopkinsville, Kentucky.

Dr. C. H. Nichols, Government Hospital for the Insane, Washington, District of Columbia.

Dr. R. J. Patterson, formerly of Hospital for the Insane, Indianapolis, Indiana.

Dr. Joseph A. Reed, Western Pennsylvania Hospital for the Insane, Pittsburg, Pennsylvania.

Dr. T. R. H. Smith, State Lunatic Asylum, Fulton, Missouri.

Dr. Nichols acted as Secretary *pro tem.*

Dr. A. McFarland was elected President, in place of Dr. Ray, who resigned.

Resolution on Dr. Ray's resignation, offered by Dr. C. H. Nichols, May 19, 1859.

Resolved, That the thanks of the Association be tendered to Dr. Ray, the late President of the Association, for his able, impartial and dignified discharge of the duties of that office.

Letters of credit were directed to be given to the British and other kindred European Associations, by the officers of the Association, to Drs. Chipley and Workman, who proposed visiting Europe.

Resolution on the death of Dr. N. Cutter, offered by Dr. G. C. S. Choate, May 17, 1859

Inasmuch as Dr. Nehemiah Cutter, an old and honored member of this Association, has deceased since our last meeting, after a long life of usefulness, nearly forty years of which was devoted to the treatment of the insane: therefore,

Resolved, That in his death we have lost a valued associate and friend, whose interest in our Association was untiring and worthy of imitation; whose zeal in the advancement of our profession continued unimpaired in advanced age, and whose genial manners and benevolent heart endeared him to all.

Several interesting papers were read; and the members were very courteously entertained by the citizens of Lexington, and after a very pleasant meeting, the Association adjourned on the 19th of June, 1859, to meet in Philadelphia.

Dr. Mount, after leaving the Hamilton County Lunatic Asylum, in 1860, resided and practiced in Cincinnati, and died in Philadelphia, February 17, 1866, from an injury received by being run over by a carriage.

The fifteenth annual meeting was held at the Continental Hotel, in the city of Philadelphia, commencing on May 28, 1860, at 10 o'clock, A. M. The following members were present:

Dr. James S. Athon, Hospital for the Insane, Indianapolis, Indiana.

Dr. D. Tilden Brown, Bloomingdale Asylum, Manhattanville, New York.

Dr. John S. Butler, Retreat for the Insane, Hartford, Connecticut.

Dr. S. W. Butler, Insane Department of the Philadelphia Almshouse.

Dr. H. A. Buttolph, State Lunatic Asylum, Trenton, New Jersey.

Dr. William A. Cheatham, Tennessee Hospital for the Insane, Nashville, Tennessee.

Dr. W. S. Chipley, Eastern Lunatic Asylum, Lexington, Kentucky.

Dr. George Cook, Brigham Hall, Canandaigua, New York.

Dr. John Curwen, Pennsylvania State Lunatic Hospital, Harrisburg, Pennsylvania.

Dr. John Fonerden, Maryland Hospital, Baltimore, Maryland.

Dr. H. M. Harlow, Hospital for the Insane, Augusta, Maine.

Dr. R. Hills, Central Ohio Lunatic Asylum, Columbus, Ohio.

Dr. O. C. Kendrick, Northern Ohio Lunatic Asylum, Newburg, Ohio.

Dr. Thomas S. Kirkbride, Pennsylvania Hospital for the Insane, Philadelphia, Pennsylvania.

Dr. Andrew McFarland, Hospital for the Insane, Jacksonville, Illinois.

Dr. J. J. McIlhenny, Southern Ohio Lunatic Asylum, Dayton, Ohio.

Dr. C. H. Nichols, Government Hospital for the Insane, Washington, District of Columbia.

Dr. William H. Prince, State Lunatic Hospital, Northampton, Massachusetts.

Dr. Isaac Ray, Butler Hospital, Providence, Rhode Island.

Dr. Joseph A. Reed, Western Pennsylvania Hospital for the Insane, Pittsburg, Pennsylvania.

Dr. W. H. Rockwell, Asylum for the Insane, Brattleboro', Vermont.

Dr. T. R. H. Smith, State Lunatic Asylum, Fulton, Missouri.

Dr. John E. Tyler, McLean Asylum, Somerville, Massachusetts.

Dr. John Waddell, Provincial Lunatic Asylum, St. John, New Brunswick.

Dr. J. H. Worthington, Friends' Asylum, Frankford, Pennsylvania.

Dr. Benjamin Ogden, Sanford Hall, Flushing, Long Island.

Dr. Edward Hall, Asylum for Criminal Insane, Auburn, New York.

Dr. J. M. Cleveland, Assistant Physician of State Lunatic Asylum, Utica, New York.

Dr. William H. Stokes, Mount Hope Institution, Baltimore, Maryland.

Dr. E. H. VanDeusen, Asylum for Insane, Kalamazoo, Michigan.

Dr. J. P. Bancroft, Lunatic Asylum, Concord, New Hampshire.

Dr. Edward R. Chapin, Kings County Lunatic Asylum, Flatbush, Long Island, New York.

Dr. L. A. Tourtellot, Assistant Physician State Lunatic Asylum, Utica, New York.

Dr. Joseph Workman, Provincial Lunatic Asylum, Toronto, Canada West.

General Allan MacDonald, of Sanford Hall.

Dr. Joseph Parrish, of Pennsylvania Training School for Feeble Minded Children, Media, Pennsylvania.

Mordecai L. Dawson and William Biddle, of the Pennsylvania Hospital for the Insane.

Dr. H. B. Wilbur, of the Asylum for Idiots, Syracuse, New York.

Dr. James Rodman, of Kentucky School for Imbecile and Idiotic Children.

Rev. Dr. Samuel Adams, Chaplain of the Eastern Kentucky Lunatic Asylum, was invited to attend the sessions of the Association.

Many interesting and valuable papers were read and discussed during this meeting, and the Association visited the Hospitals for the Insane, and a large number of public buildings in Philadelphia, and adjourned on Thursday evening, May 21, 1860, to meet in Providence, Rhode Island.

In consequence of the disturbed state of the country, caused by the breaking out of the war in April, 1861, the President sent circulars to the different members, requesting them to express to the Secretary "their views of the expediency of postponing for one year, in consideration of the disturbed state of affairs, the meeting appointed to be held in Providence, Rhode Island, on June 11, 1861."

The answer to that circular showed that twenty-one of the members favored postponement, and eight did not, so that the meeting was postponed for one year.

———o———

The sixteenth annual meeting was held at the City Hotel, Providence, Rhode Island, commencing at 10 o'clock, A. M., of June 10, 1862.

In the absence of the President and Vice President, Dr. Rockwell was chosen President *pro tem.* The following members were present :

Dr. M. Bemis, State Lunatic Hospital, Worcester, Massachusetts.

Dr. John S. Butler, Retreat for the Insane, Hartford, Connecticut.

Dr. H. A. Buttolph, State Lunatic Asylum, Trenton, New Jersey.

Dr. G. C. S. Choate, State Lunatic Hospital, Taunton, Massachusetts.

Dr. John Curwen, Pennsylvania State Lunatic Hospital, Harrisburg, Pennsylvania.

Dr. Andrew Fisher, Malden Lunatic Asylum, Amherstburg, Canada West.

Dr. John P. Gray, State Lunatic Asylum, Utica, New York.

Dr. H. M. Harlow, Hospital for the Insane, Augusta, Maine.

Dr. R. Hills, Central Ohio Lunatic Asylum, Columbus, Ohio.

Dr. O. M. Langdon, Longview Asylum, Cincinnati, Ohio,

Dr. Isaac Ray, Butler Hospital, Providence, Rhode Island.

Dr. Joseph A. Reed, Western Pennsylvania Hospital for the Insane, Pittsburgh.

Dr. W. H. Rockwell, Asylum for the Insane, Brattleboro', Vermont.

Dr. John E. Tyler, McLean Asylum, Somerville, Massachusetts.

Dr. E. H. VanDeusen, Asylum for the Insane, Kalamazoo, Michigan.

Dr. J. H. Worthington, Friends' Asylum, Philadelphia, Pennsylvania.

Dr. J. H. Woodburn, Hospital for the Insane, Indianapolis, Indiana.

Dr. Joseph Workman, Provincial Lunatic Asylum, Toronto, Canada West.

Dr. J. P. Bancroft, Asylum for the Insane, Concord, New-Hampshire.

Dr. Edward Jarvis, Dorchester, Massachusetts.

Dr. McFarland resigned the office of President, and the following officers were elected :

Dr. T. S. Kirkbride, President.

Dr. John S. Butler, Vice President.

Dr. O. M. Langdon, Treasurer.

The death of Dr. L. V. Bell was announced by Dr. Tyler, and Dr. Ray read a very admirable memoir of the life and services of Dr. Bell.

Dr. John M. Galt, for many years Superintendent of Eastern Lunatic Asylum, Williamsburg, Virginia, died on May 16, 1862.

Resolution on the Death of Dr. L. V. Bell, offered by Dr. J. E. Tyler, June 1c, 1862.

Resolved, That the members of this Association have received with emotions of profound sorrow and regret, the announcement of the death of Dr. Luther V. Bell, a past President of this body, and one of the most eminent and distinguished of the many great men who have ever adorned the medical profession. That we desire to place upon record our full and grateful appreciation of his able and un-wearied efforts and success in diffusing and establishing correct and enlightened views of the nature and treatment of mental disease; that we are deeply impressed with the remembrance of the dis-interestedness, kindness, dignity and purity of his character, of his inflexible integrity, and singular moral courage ; of his extraordin-ary attainments as a scholar, philosopher and psychologist, of the wonderful power of his personal influence, his rare and remarkable attractiveness in social life, and his inestimable worth as a friend and associate. That we recognize with unqualified admiration in all the acts of his private, professional and public life. the same unwavering consistency and faithfulness to his convictions of Tight in the face of any personal task or sacrifice which led him in the exigencies of the day, to give his life to his country, and made him. unconsciously to himself, a striking example to us all of pure, ardent, Christian pa-triotism.

Dr. John S. Butler presided at all the meetings subsequent to the first.   .

The Secretary was instructed to furnish Dr. D. T. Brown and Dr. R. Hills with a letter of introduction to Superintendents of British Institutions.

The Association were the recipients of many courtesies from the inhabitants of Providence, and were granted the opportunity 'of visiting nearly all the public buildings and institutions of dif-ferent kinds in Providence, and were also, through the courteous attention of the Trustees of the Butler Hospital, favored with an excursion to Newport, and an opportunity of seeing all the ob-jects of interest in that city.

The Association adjourned on June 10, 1862, to meet in New York.

———o———

The seventeenth annual meeting was held at the Metropolitan Hotel, in the city of New York, commencing on May 19, 1863. The following members were present :

Dr. J. P. Bancroft, Asylum for the Insane, Concord, New Hampshire.

Dr. J. W. Barstow, Sanford Hall, Flushing, Long Island.

Dr. D. Tilden Brown, Bloomingdale Asylum, New York.

Dr. John S. Butler, Retreat for the Insane, Hartford, Connecticut.

Dr. H. A. Buttolph, Lunatic Asylum, Trenton, New Jersey.

Dr. E. R. Chapin, Kings county Lunatic Asylum, Flatbush, New York.

Dr. John B. Chapin, Brigham Hall, Canandaigua, New York.

Dr. W. S. Chipley, Eastern Lunatic Asylum, Lexington, Kentucky.

Dr. J. P. Clement, Hospital for the Insane, Madison, Wisconsin.

Dr. John Curwen, Pennsylvania State Lunatic Hospital, Harrisburg, Pennsylvania.

Dr. John P. Gray, State Lunatic Asylum, Utica, New York.

Dr. Thomas S. Kirkbride, Pennsylvania Hospital for the Insane, Philadelphia.

Dr. Andrew McFarland, Hospital for the Insane, Jacksonville, Illinois.

Dr. C. H. Nichols, Government Hospital for the Insane, Washington, District of Columbia.

Dr. R. J. Patterson, Hospital for the Insane, Mt. Pleasant, Iowa.

Dr. M. H. Ranney, New York Lunatic Asylum, Blackwell's Island.

Dr. Isaac Ray, Butler Hospital, Providence, Rhode Island.

Dr. Joseph A. Reed, Western Pennsylvania Hospital for the Insane, Pittsburg, Pennsylvania.

Dr. John E. Tyler, McLean Asylum, Somerville, Massachusetts.

Dr. Clement A. Walker, Boston Lunatic Hospital, South Boston.

Dr. R. Gundry, Southern Ohio Lunatic Asylum, Dayton, Ohio.

Dr. O. M. Langdon, Longview Asylum, Cincinnati, Ohio.

Dr. J. Paregot, Yonkers, New York.

Dr. Joseph Workman, Provincial Lunatic Asylum, Toronto, Canada West.

Dr. Edward Jarvis, Dorchester, Massachusetts.

David A. Sayre, Esq., and Dr. H. M. Stillman, Trustees of the Eastern Lunatic Asylum, Lexington, Kentucky, were invited to attend the meetings of the Association.

Dr. H. B. Wilbur, of the Asylum for Idiots, Syracuse, New York, was also invited to attend the meetings.

Resolution on the Death of Dr. Hopkins, offered by Dr. R. J. Patterson, May 20, 1863.

Resolved, That in the death of Dr. R. C. Hopkins, late Superintendent of the Northern Ohio Lunatic Asylum, our specialty has lost a diligent laborer and friend, and the community in which he lived a gentleman who was true and faithful in all the relations of life.

Resolution on the Death of Drs. Morrin and Fremont, offered by Dr. J. E. Tyler, May 20, 1863.

Resolved, That we have heard, with deep regret, of the death of Drs. Morrin and Fremont, of Quebec, members of this Association, and are desirous to place upon record our sense of their great personal and professional worth, and of the great loss which we have sustained by their removal from our counsels.

The subject of a uniform law on the subject of the legal relations of the insane was introduced at this meeting by Dr. Walker, and referred to a committee, of which Dr. Ray was made chairman.

A large number of papers were read and discussed ; and the Association visited a number of institutions for the insane, and other charitable objects in and around New York, and adjourned on the 22d of May, 1863, to meet in Washington, District of Columbia.

————o————

The eighteenth annual meeting was held in Washington, District of Columbia, commencing at 10 A. M., May 10, 1864. The following members were present :

Dr. J. P. Bancroft, Asylum for the Insane, Concord, New Hampshire.

Dr. D. T. Brown, Bloomingdale Asylum, New York.

Dr. H. A. Buttolph, State Lunatic Asylum, Trenton, New Jersey.

Dr. E. R. Chapin, Kings County Lunatic Asylum, Flatbush, New York.

Dr. John Curwen, Pennsylvania State Lunatic Hospital, Harrisburg, Pennsylvania.

Dr. Pliny Earle, State Lunatic Hospital, Northampton, Massachusetts.

Dr. John P. Gray, State Lunatic Asylum, Utica, New York,

Dr. Richard Gundry, Southern Ohio Lunatic Asylum, Dayton, Ohio.

Dr. R. Hills, Central Ohio Lunatic Asylum, Columbus, Ohio.

Dr. W. P. Jones, Hospital for the Insane, Nashville, Tennessee.

Dr. T. S. Kirkbride, Pennsylvania Hospital for the Insane, Philadelphia.

Dr. J. E. J. Landry, Lunatic Asylum, Quebec, Canada East.

Dr. O. M. Langdon, Longview Asylum, Cincinnati, Ohio.

Dr. C. H. Nichols, Government Hospital for the Insane, Washington, District of Columbia.

Dr. John E. Tyler, McLean Asylum, Somerville, Massachusetts.

Dr. C. A. Walker, Lunatic Hospital, Boston, Massachusetts.

Dr. J. H. Woodburn, Hospital for the Insane, Indianapolis, Indiana.

Dr. Joshua H. Worthington, Friends' Asylum, Frankford, Pennsylvania.

Dr. E. H. Van Deusen, Asylum for the Insane, Kalamazoo, Michigan.

Dr. John Fonerden, Maryland Hospital, Baltimore.

Dr. J. D. Elbert, Trustee of the Iowa Hospital for the Insane, was invited to attend the meeting.

The Association tendered the services of the members to the Surgeon General of the United States, in view of the recent battles south of the Rappahannock river, to which the Surgeon General replied, in a very courteous note, "that should a more urgent necessity than now exists render it advisable, the offer would be gladly accepted."

The report of the chairman of the Committee on the Project of a Law determining the legal relations of the insane, was read, and the project of a law much discussed, and then postponed to a subsequent meeting.

The Association called on the President of the United States, and visited the principal buildings in and around Washington, including the Government Hospital for the Insane, and adjourned on May 13, 1864, to meet in Pittsburgh, Pennsylvania.

——o——

The nineteenth annual meeting of the Association was held in the city of Pittsburgh, Pennsylvania, commencing on June 13, 1865. The following members were present :

Dr. William S. Chipley, Eastern Lunatic Asylum, Lexington, Kentucky.

Dr. G. C. S. Choate, State Lunatic Hospital, Taunton, Massachusetts.

Dr. John Curwen, Pennsylvania State Lunatic Hospital, Harrisburg, Pennsylvania.

Dr. John Fonerden, Maryland Hospital, Baltimore, Maryland.

Dr. James R. DeWolf, Provincial Hospital for the Insane, Halifax, Novia Scotia.

Dr. Richard Gundry, Southern Ohio Lunatic Asylum, Dayton, Ohio.

Dr. R. Hills, West Virginia Hospital for Insane, Weston, West Virginia.

Dr. W. P. Jones, Hospital for the Insane, Nashville, Tennessee.

Dr. Thomas S. Kirkbride, Pennsylvania Hospital for the Insane, Philadelphia, Pennsylvania.

Dr. William L. Peck, Central Ohio Lunatic Asylum, Columbus, Ohio.

Dr. John A. Reed, Western Pennsylvania Hospital for the Insane, Pittsburg, Pennsylvania.

Dr. James Rodman, Western Lunatic Asylum, Hopkinsville, Kentucky.

Dr. John E. Tyler, McLean Asylum, Somerville, Massachusetts.

Dr. William H. Stokes, Mount Hope Institution, Baltimore, Maryland.

Dr. C. A. Walker, Lunatic Hospital, Boston, Massachusetts.

Dr. John S. Butler, Retreat for the Insane, Hartford, Connecticut.

Dr. James Douglas, Lunatic Asylum, Quebec, Canada East.

Dr. A. McFarland, Hospital for the Insane, Jacksonville, Illinois.

Dr. A. E. Kellogg, State Lunatic Asylum, Utica, New York.

Dr. H. M. Stillman, Trustee of the Eastern Lunatic Asylum

of Kentucky, and Dr. R. H. Storer, were invited to attend the meetings.

Resolution on the Death of Dr. M. H. Ranney, offered by Dr. G. C. S. Choate, June 15, 1865.

WHEREAS, It has pleased an All-Wise Providence to remove from us Dr. M. H. Ranney, late Physician and Superintendent of the New York City Lunatic Asylum, and for many years a member of this Association: therefore,

Resolved, That the intelligence of his death, in the prime of life, and at the heighth of his usefulness, has filled our hearts with sorrow; that in his devotion to his professional duties, to which he finally sacrificed his life, and in his unwavering attention to the unfortunate class under his care, we recognize a character worthy of our universal emulation; that we lament his too early decease as a loss to each of us of a warm-hearted friend and brother, to our Association of an able and valued member, and to the institution which he so long and faithfully served, of a wise and benignant head.

A number of interesting and valuable papers were read, and very interesting discussions took place on them.

Special notice was taken of a very unjust attack on Dr. L. V. Bell, in the *Journal of Mental Science*, and resolutions expressive of the sense of the Association were directed to be forwarded to the British Association.

The Association was enabled, through the attention of the Managers of the Western Pennsylvania Hospital for the Insane, to examine many of the different manufacturing establishments of Pittsburg, and other interesting objects, together with most of the public institutions of the city, and the arrangements of the Western Pennsylvania Hospital for the Insane, at Dixmont, near the city.

The Association adjourned on June 15, 1865, to meet in Washington, District of Columbia.

———o———

The twentieth annual meeting was held in the city of Washington, District of Columbia, commencing at 10½ A. M., of April 24, 1866. The following members were present :

Dr. R. Abbott, State Lunatic Asylum, Fulton, Missouri.

Dr. J. P. Bancroft, Asylum for the Insane, Concord, New Hamp-shire.

Dr. J. W. Barstow, Sanford Hall, Flushing, New York.

Dr. S. W. Butler, Insane Department of the Philadelphia Almshouse.

Dr. A. B. Cabaniss, State Lunatic Asylum, Jackson, Mississippi.

Dr. W. S. Chipley, Eastern Lunatic Asylum, Lexington, Kentucky.

Dr. George Cook, Brigham Hall, Canandaigua, New York.

Dr. John Curwen, Pennsylvania State Lunatic Hospital, Harrisburg, Pennsylvania.

Dr. James Douglas, Jr., Quebec, Canada East.

Dr.'Pliny Earle, State Lunatic Hospital, Northampton, Massachusetts.

Dr. John Fonerden, Maryland Hospital, Baltimore, Maryland.

Dr. John P. Gray, State Lunatic Asylum, Utica, New York.

Dr. W. P. Jones, Hospital for the Insane, Nashville, Tennessee.

Dr. Thomas S. Kirkbride, Pennsylvania Hospital for the Insane, Philadelphia, Pennsylvania.

Dr. Wilson Lockhart, Hospital for the Insane, Indianapolis, Indiana.

Dr. J. D. Lomax, Marshall Infirmary, Troy, New York.

Dr. C. H. Nichols, Government Hospital for the Insane, Washington, District of Columbia.

Dr. William L. Peck, Central Ohio Lunatic Asylum, Columbus, Ohio.

Dr. Mark Ranney, Hospital for Insane, Mt. Pleasant, Iowa.

Dr. J. A. Reed, Western Pennsylvania Hospital for the Insane, Dixmont, Pennsylvania.

Dr. Byron Stanton, Northern Ohio Lunatic Asylum, Newburg, Ohio.

Dr. William H. Stokes, Mount Hope Institution, Baltimore, Maryland.

Dr. John E. Tyler, McLean Asylum, Somerville, Massachusetts.

Dr. Charles E. Van Anden, Isylum for Insane Convicts, Auburn, New York.

Dr. A. H. Van Nostrand, Hospital for Insane, Madison, Wisconsin.

Dr. C. A. Walker, Lunatic Hospital, Boston, Massachusetts.

Judge Edwards, Trustee of the Iowa Hospital for the Insane, was invited to attend the meetings.

The Secretary read the correspondence between the President of the Medico-Psychological Association of Great Britain and himself, arising from the resolution relative to the attack on Dr. L. V. Bell, in the *Journal of Mental Science*.

The discussion of the Project of a Law was postponed on account of the absence of Dr. Ray.

The discussion of the proper care of the chronic insane, was continued at some length, and after the submission of several series of propositions, the following were finally agreed to :

1. Every State should make ample and suitable provision for all its insane.

2. That insane persons considered curable, and those supposed incurable, should not be provided for in separate establishments.

3. The large States should be divided into geographical districts of such size that a hospital situated at, or near, the centre of the district, will be practically accessible to all the people living within its boundaries, and available for their benefit in cases of mental disorder.

4. All State, County, and City Hospitals for the Insane, should receive all persons belonging to the vicinage designed to be accommodated by such hospital, who are affected with insanity proper, whatever may be the form, or nature, of the bodly disease accompanying the mental disorder.

5. All hospitals for the insane should be constructed, organized and managed, substantially in accordance with the propositions adopted by the Association in 1851 and 1852, and still in force.

6. The facilities for classification, or ward separation, possessed by each institution, should equal the requirements of the different conditions of the several classes received by such institutions whether those different conditions are mental or physical in their character.

7. The enlargement of a city, county or State institution for the insane which, in the extent and character of the district in which it is situated, is conveniently accessible to all the people of such district, may be properly carried, as required, to the extent of accommodating six hundred patients, embracing the usual proportions of curable and incurable insane in a particular community.

These propositions were unanimously adopted, except the last, and on that the vote stood eight in the affirmative and six in tl.e negative. Yeas—Abbot, Cabaniss, Chipley, Earle, Gray, Lomax, Nichols and Van Nostrand. Nays—Cook, Curwen, Jones, Kirkbride, Lockhart and Walker, and on the final adoption of the resolutions, the affirmative votes were Abbot, Cabaniss, Chipley, Earle, Gray, Lockhart, Lomax, Nichols and Van Nostrand.

The negative votes were Cook, Curwen, Jones, Kirkbride and Walker.

The Association visited the President of the United States, the Army Medical Museum, and the Government Hospital for Insane, and also several other buildings in the city, and adjourned on April 27, 1866, to meet in Philadelphia.

————o————

The twenty-first annual meeting was held in the city of Philadelphia, commencing at 10 o'clock A. M., on May 21, 1867. The following members were present :

Dr. J. P. Bancroft, Asylum for the Insane, Concord, New-Hampshire.

Dr. William P. Beall, State Lunatic Asylum, Austin, Texas.

Dr. D. T. Brown, Bloomingdale Asylum, New York.

Dr. H. A. Buttolph, State Lunatic Asylum, Trenton, New Jersey.

Dr. J. P. Chapin, Brigham Hall, Canandaigua, New York.

Dr. John Curwen, Pennsylvania State Lunatic Hospital, Harrisburg, Pennsylvania.

Dr. Pliny Earle, State Lunatic Hospital, Northampton, Massachusetts.

Dr. Edward C. Fisher, Asylum for the Insane, Raleigh, North Carolina.

Dr. John Fonerden, Maryland Hospital, Baltimore, Maryland.

Dr. Richard Gundry, Southern Ohio Lunatic Asylum, Dayton, Ohio.

Dr. R. Hills, Hospital for the Insane, Weston, West Virginia.

Dr. Thomas S. Kirkbride, Pennsylvania Hospital for the Insane, Philadelphia.

Dr. J. E. J. Landry, Lunatic Asylum, Quebec, Canada East.

Dr. Andrew McFarland, Hospital for the Insane, Jacksonville' Illinois.

Dr. C. H. Nichols, Government Hospital for the Insane, Washington, District of Columbia.

Dr. R. L. Parsons, New York City Lunatic Asylum, New York.

Dr. William L. Peck, Central Ohio Lunatic Asylum, Columbus, Ohio.

Dr. Isaac Ray, Providence, Rhode Island.

Dr. D. D. Richardson, Insane Department of the Philadelphia Almshouse.

, Dr. James Rodman, Western Lunatic Asylum, Hopkinsville, Kentucky.

Dr. Byron Stanton, Northern Ohio Lunatic Asylum, Newburg, Ohio.

Dr. L. A. Tourtellot, First Assistant Physician of State Lunatic Asylum, Utica, New York.

Dr. C. A. Walker, Lunatic Hospital, Boston, Massachusetts.

Dr. Benjamin Workman, Assistant Medical Superintendent of Provincial Lunatic Asylum, Toronto, Canada West.

Dr. J. H. Worthington, Friends' Asylum, Philadelphia, Pennsylvania.

Dr. J. A. Reed, Western Pennsylvania Hospital for the Insane, Dixmont, Pennsylvania.

Dr. E. R. Chapin, Kings County Lunatic Asylum, Flatbush, New York.

Dr. A. B. Cabaniss, Lunatic Asylum, Jackson, Mississippi.

Dr. J. D. Lomax, Marshall Infirmary, Troy, New York.

Dr. Charles E. Van Anden, Asylum for Insane Convicts, Auburn, New York.

Dr. Edward Jarvis, Dorchester, Massachusetts.

Dr. Charles H. Hughes, State Lunatic Asylum, Fulton, Missouri.

Dr. George Brown, Asylum for Idiots, Barre, Massachusetts, and Dr. H. B. Wilbur, Asylum for Idiots, Syracuse, New York, were invited to attend the meetings. Dr. William B. Atkinson, Secretary of the American Medical Association, and Dr. John Hart, of New York, were also invited to attend the meetings.

The following resolutions were adopted in reference to the proceedings of the Association.

1. *Resolved,* That for the present meeting and in the future, it be the duty of the Secretary to secure a phonographic report of the proceedings of the Association.

2. That after each annual meeting, he shall forward a copy of said report for insertion in the *Journal of Insanity,* provided that, before forwarding it for publication, every member shall have the opportunity to revise his reported remarks, and after its publication shall be supplied, at his own expense for paper and press work, with such number of pamphlet copies of the whole report as he may order.

3. That in the revision of remarks, verbal alterations alone shall

be permitted. No new matter further than this shall be introduced, but all or any parts of the matter as reported may be suppressed or condensed at the discretion of the Secretary.

4. The report shall be published, if published at all, as furnished by the Secretary.

5. That the expense of reporting the proceedings, and preparing them for publication, be defrayed by an annual assessment upon the members sufficient for the purpose.

The Association visited in particular, the Pennsylvania Hospital for the Insane, the Friends' Asylum, and the Insane Department of the Philadelphia Almshouse, and for want of time were compelled to decline many invitations to visit various institutions in the city.

The Association adjourned on May 25, 1867, to meet in Boston, Massachusetts.

———o———

The twenty-second annual meeting was held at the American House, in the city of Boston, commencing on June 2, 1868. The following members were present :

Dr. J. P. Bancroft, New Hampshire Asylum for the Insane, Concord, New Hampshire.

Dr. J. W. Barstow, Sanford Hall, Flushing, New York.

Dr. John S. Butler, Retreat for the Insane, Hartford, Connecticut.

Dr. H. A. Buttolph, State Lunatic Asylum, Trenton, New Jersey.

Dr. Edward R. Chapin, Kings County Lunatic Asylum, Flatbush, Long Island, New York.

Dr. W. S. Chipley, Eastern Lunatic Asylum, Lexington, Kentucky.

Dr. G. C. S. Choate, Taunton Lunatic Hospital, Taunton, Massachusetts.

Dr. John Curwen, Pennsylvania State Lunatic Hospital, Harrisburg, Pennsylvania.

Dr. Pliny Earle, Northampton Lunatic Asylum, Northampton, Massachusetts.

Dr. H. M. Harlow, Hospital for the Insane, Augusta, Maine.

Dr. R. Hills, West Virginia Hospital for the Insane, Weston, West Virginia.

Dr. C. H. Hughes, State Lunatic Asylum, Fulton, Missouri.

Dr. W. P. Jones, Hospital for the Insane, Nashville, Tennessee.

Dr. Thomas S. Kirkbride, Pennsylvania Hospital for the Insane, Philadelphia.

Dr. C. H. Nichols, Government Hospital for the Insane, Washington, District of Columbia.

Dr. R. L. Parsons, City Lunatic Asylum, New York City.

Dr. Isaac Ray, Philadelphia, Pa.

Dr. Mark Ranney, Iowa Hospital for the Insane, Mt. Pleasant, Iowa.

Dr. D. D. Richardson, Insane Department Philadelphia Almshouse.

Dr. John W. Sawyer, Butler Hospital, Providence, Rhode Island.

Dr. S. S. Schultz, State Hospital for the Insane, Danville, Pennsylvania.

Dr. Samuel E. Shantz, Minnesota Hospital for the Insane, St. Peter, Minnesota.

Dr. A. Marvin Shew, General Hospital for the Insane, Middletown, Connecticut.

Dr. Byron Stanton, Northern Ohio Lunatic Asylum, Newburg, Ohio.

Dr. F. T. Stribling, Western Lunatic Asylum, Staunton, Virginia.

Dr. John E. Tyler, McLean Asylum, Somerville, Massachusetts.

Dr. C. A. Walker, Boston Lunatic Hospital, South Boston, Massachusetts.

Dr. Jos. Draper. Worcester Lunatic Hospital, Worcester, Massachusetts.

Dr. R. Gundry, Southern Ohio Lunatic Asylum, Dayton, Ohio.

Dr. Edward Jarvis, Dorchester, Massachusetts.

Dr. W. Lockhart, Hospital for the Insane, Indianapolis, Indiana.

Dr. Joseph D. Lomax, Marshall Infirmary, Troy, New York.

Dr. Charles A. Lee, Delegate of the American Medical Association,

Dr. A. W. McClure, Trustee of Iowa Hospital for the Insane.

Samuel D. Sewall, Trustee of the Worcester Lunatic Hospital, Massachusetts.

Dr. Henry L. Sabin, Trustee of the Lunatic Hospital, Northampton, Massachusetts.

Dr. George Brown, Asylum for Idiots, Barre, Massachusetts.

The principal business of the Association was the discussion and adoption of a Project of Law for determining the legal relations of the Insane.

The Association also adopted a memorial to the Congress of the United States in favor of relieving from political disabilities the Superintendents of the Hospitals for the Insane in the States lately in rebellion.

Dr. C. A. Walker read a memoir of the late Dr. Charles H. Stedman.

The Association visited all the institutions for the insane in Boston, and also several of the Hospitals for general diseases, and other institutions, and received many courtesies from the officers and trustees of the principal institutions for the insane, and also the officers of Harvard College.

Samuel E. Shantz, M. D., died at St. Peter, Minnesota, on August 22, 1868.

He was born in Waterloo township, Canada, received his education at the University of Toronto and at Harvard, where he finished his medical course ; was for several years a surgeon in the army in Virginia during the rebellion, and afterwards Second Assistant Physician in the New York State Lunatic Asylum, at

Utica, whence he was called in 1866, to become Superintendent of the Minesota State Hospital for Insane at St. Peter.

He was engaged in directing the erection of the hospital at St. Peter, when he was attacked with typhoid fever, and died in the fourth week of his sickness.

"He was a member of the Episcopal Church. He had married an accomplished and most estimable lady of Utica, New York, only about three months previous to his death. His life was gentle and pure, and his end was peace."

History of the Project of the Law for regulating the Legal Relations of the Insane, recommended by the Association of Medical Superintendents of American Institutions for the Insane.

The very serious deficiencies in the existing laws respecting insanity and the insane, have frequently been the subject of discussion, in the meetings of the "Association of Medical Superintendents of American Institutions for the Insane," and the necessity acknowledged of some legislation that should fully meet the requirements of the case. It became a prevalent sentiment that it was incumbent on the Association to publish its views on this subject, for the reason that no other class of persons is so well acquainted with the consequences of the present legislation, or, rather, want of legislation, or so well fitted by its habitual pursuits, to suggest the appropriate measures. It must be regarded as a fundamental principle, that the laws respecting insanity should be in accordance with the present state of our knowledge concerning it, which is far more ample and certain than that which existed two hundred years ago and still continues to determine, more or less, the opinions of legislators and Judges. In respect to some of the relations of the insane, even that of their admission into and discharge from hospitals and asylums, the law in some States is entirely silent ; and in respect to some others, the law is not in accordance with the most enlightened views of the disease, or the best welfare of the insane. As might be supposed, this state of things has occasioned, of late years, much public dissatisfaction, one of the worst effects of which has been to debar many from enjoying the benefits of those institutions which have proved to be the best adapted, of all existing instrumentalities, for the cure, or custody, of the insane.

At the meeting of the Association in New York, May, 1863, a committee was appointed to examine the whole subject and make special inquiry as to the legislation which the case requires. The committee consisted of one from each State, viz: Drs. Harlow, of Maine ; Bancroft, of New Hampshire ; Rockwell, of Vermont ; Jarvis, of Massachusetts ; Ray, of Rhode Island ; Butler, of Connecticut ; Gray, of New York ; Buttolph, of New Jersey ; Curwen, of Pennsylvania ; Fonerden, of Maryland ; Nichols, of the District of Columbia ; Gundry, of Ohio ; Woodburn of Indiana ; McFarland, of Illinois ; Van Deusen, of Michigan ; Clement, of Wisconsin ; Patterson, of Iowa ; Smith, of Missouri ; Chipley, of Kentucky; Jones, of Tennessee, and Workman, of Canada and the British Provinces. At the next meeting of the Association, in Washington, in May, 1864, the Committee reported, through its chairman, Dr. Ray, which report was accompanied by the Project of a General Law for regulating the most important relations of the Insane. After considerable discussion, the further consideration of the subject was postponed to the next meeting, but owing to the absence of the chairman, it was not resumed until the last meeting of the Association, in Boston, June, 1868. Then it was again most thoroughly discussed ; the various sections—most of them more or less modified—adopted with little dissent, one after another, and finally adopted as a whole, with the accompanying preamble, unanimously.

It may be well to anticipate an objection that may be made to one feature of this project, viz : that the Association has gone beyond its proper province in prescribing the rule of law applicable to certain cases, and thereby usurping the functions of the lawyer. The objection assumes—what we are not willing to grant—that these cases are exclusively questions of law, to be determined without any reference to their medical aspects. If

insanity is a disease, the laws respecting it must be framed in due
accordance with its influence on the mental condition, as observed by medical men. Framed upon any other principle, they
can only be arbitrary and capricious, reflecting the learning of
the past, or the current notions of the present, and consequently
subject to change and uncertainty. If the opinions of medical
men, on questions of insanity, are entitled to any weight whatever, they cannot be restricted to this or that particular point,
but must be received for what they are worth, wherever they can
be supposed to throw any light on the mental condition. Therefore, the law respecting the effect of insanity on wills, and contracts, should reflect our actual knowledge of the disease no less
than that respecting its effect on criminal acts. It cannot be
denied that, under the rules of law accepted in our courts, de-
cisions on these subjects have been rendered greatly at variance
with those views of mental disease which have resulted from the
larger observation, and more exact inquiry of our own time.
The Association has only acted upon the self-evident principle
that the effect of insanity on the mental operations, is a professional question, whether it has reference to criminal, or civil acts.

The Project of a General Law, as finally adopted, was endorsed
by every member of the Association then present, and therefore
is free from any distrust that might have attached to it, had it
been adopted by a bare majority of votes. Embodying, as it
does, only those conclusions in regard to which there could be
no diversity of opinion, there is no presumption in claiming for
it an authority that ought to be felt in all future legislation on
the subject. If it be objected that the peculiar vocation of the
members gave them a bias against the conclusions of those who
regard the subject from other and very different points of view,
the Association is ready, no doubt, to acknowledge the correctness of the objection, if it means that bias which springs from
extraordinary opportunities for observing the consequences of
the present legal deficiencies, and from that habit of mind and
pursuit, which best enables one to devise the appropriate remedy
for a practical evil. For it must be considered that the professional experience of most of these men extends over a period of

many years—of a quarter of a century, or more, perhaps—and that the circumstances of their calling have made them well acquainted with the opinions and feelings of the sane, as well as the insane, with the grievances suffered by both, and the requirements that both demand. Any bias, therefore, which results from superior knowledge, should be welcomed, rather than made a matter of reproach, unless we adopt the principle that in all inquiries after the truth, the knowledge of one man may be fairly offset by the ignorance of another.

In preparing the following legal provisions, the Association has aimed at such an adjustment of the rights, and duties, the abilities and disabilities, both of the insane and of all others, directly, or indirectly connected with them, as is consistent with exact justice to all, and the highest welfare of the insane, avoiding, if possible, on the one hand, the charge of excessive indulgence towards the insane, and on the other of unduly strengthening and extending the control of the family, or the public. There is a small class of persons in the community to whom the conclusions of the Association will be highly unsatisfactory—to whom the alleged grievances seem to require extreme remedies, meaning, thereby, measures that, while they prevent a contingent and very limited evil, inflict a positive harm, and one of indefinite extent. It has seemed to the Association, however, that all sound legislation should be directed towards the common, not the exceptional cases, and affect a greatly preponderating balance of good. Besides, the trouble in question cannot be reached by statutes, or legal processes. Positive wrongs may be abated by legislation, but not so that popular sensitiveness which springs from ignorance, prejudice, unreasonable suspicion, or constitutional distrust. It can be removed only by a removal of the cause in which it originates.

With these introductory remarks, the Association submits the accompanying Project of a Law, in the hope that to every intelligent and unprejudiced mind, it will appear well calculated to accomplish the proposed object in the manner most consistent with the teachings of science, the demands of justice, and the claims of humanity.

Project of the Law.

The Association of Medical Superintendents of American Institutions for the Insane, believing that certain relations of the
insane should be regulated by statutory enactments calculated to
secure their rights and also the rights of those entrusted with
their care, or connected with them by ties of relation, or friendship, as well as to promote the ends of justice, and enforce the
claims of an enlightened humanity, for this purpose recommend
that the following legal provisions be adopted by every State
whose existing laws do not, already, satisfactorily provide for
these great ends :

1. Insane persons may be placed in a hospital for the insane
by their legal guardians, or by their relatives, or friends, in case
they have no guardians ; but never without the certificate of one
or more reputable physicians, after a personal examination, made
within one week of the date thereof; and this certificate to be
duly acknowledged before some magistrate, or judicial officer,
who shall certify to the genuineness of the signature, and to the
respectability of the signer.

2. Insane persons may be placed in a hospital, or other suitable place of detention, by order of a magistrate, who, after
proper inquisition, shall find that such persons are at large, and
dangerous to themselves or others, or require hospital care and
treatment, while the fact of their insanity shall be certified by
one, or more, reputable physicians, as specified in the preceding
section.

3. Insane persons may be placed in a hospital, by order of any
high judicial officer, after the following course of proceedings,
viz: on statement in writing, of any respectable person, that a
certain person is insane, and that the welfare of himself, or of
others, requires his restraint, it shall be the duty of the judge to
appoint, immediately, a commission, who shall inquire into and
report upon, the facts of the case. If, in their opinion, it is a
suitable case for confinement, the judge shall issue his warrant
for such disposition of the insane person as will secure the objects of the measure.

4. The commission provided for in the last section, shall be composed of not less than three nor more than four persons, one of whom, at least, shall be a physician and another a lawyer. In their inquisition they shall hear such evidence as may be offered touching the merits of the case, as well as the statements of the party complained of, or of his counsel. The party shall have seasonable notice of the proceedings, and the judge is authorized to have him placed in suitable custody while the inquisition is pending.

5. On a written statement being addressed, by some respectable person, to any high judicial officer, that a certain person, then confined in a hospital for the insane, is not insane, snd is thus unjustly deprived of his liberty, the judge, at his descretion, shall appoint a commission of not less than three, nor more than four, persons, one of whom, at least, shall be a physician, and another a lawyer, who shall hear such evidence as may be offered touching the merits of the case, and, without summoning the party to meet them, shall have a personal interview with him, so managed as to prevent him, if possible, from suspecting its objects. They shall report their proceedings to the judge, and if, in their opinion, the party is not insane, the judge shall issue an order for his discharge.

6. If the officers of any hospital shall wish for a judicial examination of a person in their charge, such examination shall be had in the manner provided in the fifth section.

7. The commission provided for in the fifth section shall not be repeated, in regard to the same party, oftener than once in six months ; and in regard to those placed in a hospital under the third section, such commission shall not be appointed within the first six months of their residence therein.

8. Persons placed in a hospital under the first section of this act, may be removed therefrom by the party who placed them in it.

9. Persons placed in a hospital under the second section of this act, may be discharged by the authorities in whom the government of the hospital is vested.

10. All persons whose legal status is that of paupers, may be placed in a hospital for the insane, by the municipal authorities who have charge of them, and may be removed by the same authority, the fact of insanity being established as in the first section.

11. On statement, in writing, to any high judicial officer, by some friend of the party, that a certain party, placed in a hospital under the third section, is losing his bodily health, and that consequently his welfare would be promoted by his discharge; or that his mental disease has so far changed its character as to render his further confinement unnecessary, the judge shall make suitable inquisition into the merits of the case, and according to its result, may or may not, order the discharge of the party.

12. Persons placed in any hospital for the insane, may be removed therefrom, by parties who have become responsible for the payment of their expenses; provided that such obligation was the result of their own free act and accord, and not of the operation of law, and that its terms require the removal of the patient in order to avoid further responsibility.

13. Insane persons shall not be made responsible for criminal acts in a criminal suit, unless such acts shall be proved not to have been the result, directly, or indirectly, of insanity.

14. Insane persons shall not be tried for any criminal act during the existence of their insanity; and for settling this issue, one of the judges of the court by which the party is to be tried, shall appoint a commission, consisting of not less than three, nor more than five, persons, all of whom shall be physicians, and one, at least, if possible, an expert in insanity, who shall examine the accused, hear the evidence that may be offered touching the case, and report their proceedings to the judge, with their opinions respecting his mental condition. If it be their opinion that he is not insane, he shall be brought to trial; but if they consider him insane, or are in doubt respecting his mental condition, the judge shall order him to be placed in some hospital for the insane, or some other place favorable for a scientific observation of his mental condition. The person to whose custody he may be

committed, shall report to the judge respecting his mental condition, previous to the next term of court ; and if such report is not satisfactory, the judge shall appoint a commission of inquiry, in the manner just mentioned, whose opinion shall be followed by the same. proceedings as in the first instance.

15. Whenever any person is acquitted, in a criminal suit, on the ground of insanity, the jury shall declare this fact in their verdict ; and the court shall order the prisoner to be committed to some place of confinement, for safe keeping, or treatment, there to be retained until he may be discharged in the manner provided in the next section.

16. If any judge of the highest court having original jurisdiction, shall be satisfied, by the evidence presented to him, that the prisoner has recovered, and that the paroxysm in insanity in which the criminal act was committed, was the first and only one he had ever experienced, he may order his unconditional discharge ; if, however, it shall appear that such paroxysm of insanity was preceded by at least one other, then the court may, in its discretion, appoint a guardian of his person, and to him commit the care of the prisoner, said guardian giving bonds for any damage his ward may commit : *Provided, always,* That in case of homicide, or attempted homicide, the prisoner shall not be discharged, unless by the unanimous consent of the Superintendent and the managers of the hospital, and the court before which he was tried.

17. If it shall be made to appear to any judge of the supreme judicial court, or other high judicial officer, that a certain insane person is manifestly suffering from the want of proper care, or treatment, he shall order such person to be placed in some hospital for the insane, at the expense of those who are legally bound to maintain them.

18. Application for the guardianship of an insane person shall be made to the judge of probate, or judge having similar jurisdiction, who, after a hearing of the parties, shall grant the measure, if satisfied that the person is insane, and incapable of managing his affairs discreetly. Seasonable notice shall be

given to the person who is the object of the measure, if at large, and if under restraint, to those having charge of him ; but his presence in court, as well as the reading of the notice to him, may be dispensed with, if the court is satisfied that such reading, or personal attendance, would probably be detrimental to his mental, or bodily health. The removal of the guardianship shall be subjected to the same mode of procedure as its appointment.

19. Insane persons shall be made responsible, in a civil suit, for any injury they may commit upon the person, or property of others ; reference being had in regard to the amount of damages, to the pecuniary means of both parties, to the provocation sustained by the defendant, and any other circumstance which, in a criminal suit, would furnish ground for mitigation of punishment.

20. The contracts of the insane shall not be valid, unless it can be shown, either that such acts were for articles of necessity, or comfort, suitable to the means and condition of the party, or that the other party had no reason to suspect the existence of any mental impairment and that the transaction exhibited no marks of unfair advantage.

21. A will may be invalidated on the ground of the testator's insanity, provided it be proved that he was incapable of understanding the nature and consequences of the transaction, or of appreciating the relative values of property, or of remembering and calling to mind all the heirs-at-law, or of resisting all attempts to substitute the will of others for his own. A will may also be invalidated on the ground of the testator's insanity, provided it be proved that he entertained delusions respecting any heirs-at-law, calculated to produce unfriendly feeling towards them.

———o———

The twenty-third annual meeting was held at Staunton, Virginia, commencing on June 15, 1869. The following members were present :

Dr. D. R. Brower, Eastern Lunatic Asylum, Williamsburg, Virginia.

Dr. D. Tilden Brown, Bloomingdale Asylum, New York City.

Dr. John S. Butler, Retreat for the Insane, Hartford, Connecticut.

Dr. A. B. Cabaniss, State Lunatic Asylum, Jackson, Mississippi.

Dr. Edward R. Chapin, Kings County Lunatic Asylum, Flatbush. New York.

Dr. John Curwen, Pennsylvania State Lunatic Hospital, Harrisburg, Pennsylvania.

Dr. F. T. Fuller, Assistant Physician Insane Asylum, Raleigh, North Carolina.

Dr. B. Graham, State Lunatic Asylum, Austin, Texas.

Dr. John P. Gray, State Lunatic Asylum, Utica, New York.

Dr. Edward Jarvis, Dorchester, Massachusetts.

Dr. Henry Landor, Malden Asylum, Amherstburg, Ontario.

Dr. Alexander S. McDill, State Hospital for the Insane, Madison, Wisconsin.

Dr. Edward Mead, Cincinnati, Ohio.

Dr. Charles H. Nichols, Government Hospital for the Insane, Washington, District of Columbia.

Dr. R. Hills, Hospital for the Insane, Weston, West Virginia.

Dr. C. H. Hughes, State Lunatic Asylum, Fulton, Missouri.

Dr. W. P. Jones, Hospital for the Insane, Nashville, Tennessee.

Dr. Thomas S. Kirkbride, Pennsylvania Hospital for the Insane, Philadelphia, Pennsylvania.

Dr. Isaac Ray, Philadelphia, Pennsylvania.

Dr. James Rodman, Western Lunatic Asylum, Hopkinsville, Kentucky.

Dr. Francis T. Stribling, Western Lunatic Asylum, Staunton, Virginia.

Dr. A. M. Shew, General Hospital for the Insane. Middletown, Connecticut.

Dr. John E. Tyler, McLean Asylum, Somerville, Massachusetts.

Dr. C. A. Walker, Lunatic Hospital, Boston, Massachusetts.

Dr. Joseph Workman, Provincial Lunatic Asylum, Toronto. Ontario.

Dr. Walker, after a few remarks in reference to the cause of the death of Dr. Fonerden, offered the following resolutions, which were unanimously adopted:

Resolved, That in the death of Dr. John Fonerden, Superintendent of the Maryland Hospital, Baltimore, this Association has lost one of its early and valued members, the cause a tried and faithful supporter, the community a Christian gentleman, and ourselves a genial and true-hearted friend.

Resolved, That we sympathize with the managers of the Maryland Hospital for the Insane, in the loss of their devoted, long-serving and judicious superintendent.

Resolved, That our hearts ache for his stricken family in their sudden and great bereavement.

Dr. Gray offered the following resolution in regard to the death of Dr. Shantz :

WHEREAS since the last meeting of this Association, Dr. Samuel E. Shantz, Superintendent of the Minnesota Hospital for the Insane one of its members has been called away by death ; therefore,

Resolved, That, while lamenting his early death, and while recognizing in the sad event the hand of God, whose ways are not as man's ways, and who alone doeth all things well, we desire to express and record our sense of the loss to the Medical Profession and to this Association of a young man of promise at the very outset of a career of honor and usefulness.

Resolved, That we hereby tender to his early bereaved wife and to his family our profound sympathy in their deep affliction, and that the Secretary of the Association be directed to transmit to Mrs. Shantz and to the family of our late associate, a copy of these resolutions.

These were unanimously adopted.

Dr. Workman offered the following resolution on the death of Dr. I. P. Litchfield, which was unanimously adopted :

Resolved, That this Association, having learned of the death of Dr.

I. P. Litchfield, Superintendent of the Rockwood Asylum, Canada West, desires to record its appreciation of the valuable administrative qualities evinced by him in the discharge of his official duties, and to express to his widow, its sincere condolence in the bereavement to which she has been subjected by this dispensation of Providence.

Dr. Stribling was appointed to prepare a biographical sketch of Dr. Fonerden ; Dr. Workman of Dr. Shantz, and Dr. Landor of Dr. Litchfield.

An invitation was received from the Association to attend the laying of the corner stone of the State Hospital for the Insane, at Danville, Pennsylvania, on August 26, 1869.

Dr. Robert Reyburn, delegate of the American Medical Association, and Surgeon John Moore, United States Army, President of the Board of Directors of the Eastern Lunatic Asylum, Williamsburg, Virginia, were introduced by the President.

The following resolution in regard to religious services in institutions for the insane was adopted :

Resolved, That this Association hereby expresses its earnest conviction that religious services of some kind in our institutions for the insane are generally highly salutary to their inmates, and should be regularly held, and that the Association hereby reaffirms the ninth proposition of the series adopted in relation to the organization and management of hospitals for the insane in 1856.

The place of next meeting was selected at Hartford, Connecticut, and the time fixed for the third Wednesday of June, 1870.

Dr. Nichols presented to the Association the project of a system of statistics adopted at the International Congress of Alienists, held in Paris in 1867, and the papers were referred to a committee consisting of Drs. Jarvis, Nichols and Stribling.

———o———

The twenty-fourth annual meeting was held at Hartford, Connecticut, commencing on June 15, 1870. The following members were present :

Dr. J. P. Bancroft, Asylum for the Insane, Concord, New-Hampshire.

Dr. J. W. Barstow, Sanford Hall, Flushing, New York.

Dr. D. R. Brower, Eastern Lunatic Asylum, Williamsburg, Virginia.

Dr. D. Tilden Brown, Bloomingdale Asylum, New York.

Dr. Henry W. Buel, Spring Hill Institution, Litchfield, Connecticut.

Dr. John S. Butler, Retreat for the Insane, Hartford, Connecticut.

Dr. H. A. Buttolph, State Lunatic Asylum, Trenton, New Jersey.

Dr. John H. Callender, Hospital for the Insane, Nashville, Tennessee.

Dr. E. R. Chapin, Kings County Lunatic Asylum, Flatbush, Long Island, New York.

Dr. George C. S. Choate, New York.

Dr. John Curwen, State Lunatic Hospital, Harrisburg, Pennsylvania.

Dr. James R. DeWolf, Hospital for the Insane, Halifax, Nova Scotia.

Dr. J. P. Dudley, Assistant Physician Eastern Lunatic Asylum, Lexington, Kentucky.

Dr. Pliny Earle, Lunatic Hospital, Northampton, Massachusetts.

Dr. Orpheus Everts, Hospital for the Insane, Indianapolis, Indiana.

Dr. W. W. Godding, Lunatic Hospital, Taunton, Massachusetts.

Dr. John P. Gray, State Lunatic Asylum, Utica, New York.

Dr. Thomas F. Greene, State Lunatic Asylum, Milledgeville, Georgia.

Dr. Eugene Grissom, Insane Asylum, Raleigh, North Carolina.

Dr. Richard Gundry, Southern Ohio Lunatic Asylum, Dayton, Ohio.

Dr. Henry M. Harlow, Hospital for the Insane, Augusta, Maine.

Dr. R. Hills, Hospital for Insane, Weston, West Virginia.

Dr. Edward Jarvis, Dorchester, Massachusetts.

Dr. Thomas S. Kirkbride, Pennsylvania Hospital for the Insane, Philadelphia. Pennsylvania.

Dr. J. M. Lewis, Northern Ohio Lunatic Asylum, Newburg, Ohio.

Dr. Alexander S. McDill, Hospital for the Insane, Madison, Wisconsin.

Dr. C. H. Nichols, Government Hospital for the Insane, Washington, District of Columbia.

Dr. Ralph L. Parsons, New York City Lunatic Asylum.

Dr. William Porter, Spring Hill Institution, Litchfield, Connecticut.

Dr. Mark Ranney, Hospital for Insane, Mt. Pleasant, Iowa.

Dr. Isaac Ray, Philadelphia, Pennsylvania.

Dr. Joseph A. Reed, Western Pennsylvania Hospital for the Insane, Dixmont, Allegheny County, Pennsylvania.

Dr. D. D. Richardson, Department for the Insane, Philadelphia Hospital.

Dr. William H. Rockwell, Asylum for the Insane, Brattleboro', Vermont.

Dr. John W. Sawyer, Butler Hospital, Providence, Rhode Island.

Dr. A. M. Shew, General Hospital for the Insane, Middletown, Connecticut.

Dr. William F. Steuart, Maryland Hospital, Baltimore.

Dr. C. A. Walker, Lunatic Hospital, Boston, Massachusetts.

Dr. J. H. Worthington, Friends' Asylum for the Insane, Frankford, Pennsylvania.

Also the following gentlemen by invitation :

Dr. John L. Atlee, Delegate of the American Medical Association.

Dr. Wilmer Worthington, General Agent and Secretary of the Board of Public Charities of Pennsylvania.

Dr. James P. White, President of the Board of Managers of the Buffalo State Asylum for the Insane.

Dr. H. B. Wilbur, Asylum for Idiots, Syracuse, New York.

Dr. George Brown, Institution for Feeble Minded Youth, Barre, Massachusetts.

Dr. H. M. Knight, School for Imbeciles, Lakeville, Connecticut.

Dr. J. H. Woodburn, Commissioner of Insane Hospital, Indianapolis, Indiana.

Dr. E. T. Elkins, Commissioner of Insanity for California.

Luke Palmer, Trustee of Iowa Hospital for the Insane.

L. F. Boughton, President of Board of Trustees of State Lunatic Asylum, Milledgeville, Georgia.

Frederick H. Wines, Secretary of Board of Charities of Illinois.

Dr. G. Seguin, New York.

Dr. E. C. Seguin, New York City.

Dr. Kirkbride resigned the office of President, and Dr. John S. Butler was chosen President; Dr. Charles H. Nichols, Vice President, and Dr. John Curwen, Secretary and Treasurer.

Dr. John Curwen was appointed delegate to the American Medical Association to be held in San Francisco, and Dr. G. A. Shurtleff, Alternate.

Dr. J. W. Barstow read a sketch of the life of Dr. Edward Hall, late of the Asylum for Criminal Insane at Auburn, New York, who died at Messina, Sicily, on April 28, 1870.

Toronto, Canada, was selected as the next place of meeting, and the first Tuesday of June, 1871, as the time.

Dr. Jarvis presented the report of the Committee on Statistics, which was, on motion, made the special order for the next meeting of the Association in 1871.

The Association visited the Retreat for the Insane at Hartford, and also the General Hospital for the Insane at Middletown.

The twenty-fifth annual meeting was held in Toronto, Ontario, commencing on June 6, 1871. The following members were present:

Dr. J. P. Bancroft, Asylum for the Insane, Concord, New Hampshire.

Dr. D. T. Brown, Bloomingdale Asylum, New York City.

Dr. John S. Butler, Retreat for the Insane, Hartford, Connecticut.

Dr. T. B. Camden, Superintendent elect Hospital for the Insane, Weston, West Virginia.

Dr. John Clopton, Assistant Physician, Eastern Lunatic Asylum, Williamsburg, Virginia.

Dr. William M. Compton, State Lunatic Asylum, Jackson, Mississippi.

Dr. D. B. Conrad, Central Lunatic Asylum, near Richmond, Virginia.

Dr. George Cook, Brigham Hall, Canandaigua, New York.

Dr. John Curwen, Pennsylvania State Lunatic Hospital, Harrisburg, Pennsylvania.

Dr. James. R. DeWolf, Provincial Hospital for the Insane, Halifax, Nova Scotia.

Dr. John R. Dickinson, Kingston Asylum, Ontario, Canada.

Dr. J. F. Ensor, Lunatic Asylum, Columbia, South Carolina.

Dr. Orpheus Everts, Hospital for the Insane, Indianapolis, Indiana.

Dr. John P. Gray, State Lunatic Asylum, Utica, New York.

Dr. Eugene Grissom, Asylum for the Insane, Raleigh, North Carolina.

Dr. Richard Gundry, Southern Ohio Lunatic Asylum, Dayton, Ohio.

Dr. C. H. Hughes, State Lunatic Asylum, Fulton, Missouri.

Dr. Edward Jarvis, Dorchester, Massachusetts.

Dr. T. S. Kirkbride, Pennsylvania Hospital for the Insane, Philadelphia.

Dr. Henry Landor, London Asylum, Ontario, Canada.

Dr. J. M. Lewis, Northern Ohio Lunatic Asylum, Newburg, Ohio.

88

Dr. Joseph D. Lomax, Marshall Infirmary, Troy, New York.

Dr. A. S. McDill, Hospital for the Insane, Madison, Wisconsin.

Dr. A. E. McDonald, Assistant Physician, Lunatic Asylum, Ward's Island, New York.

Dr. Charles H. Nichols, Government Hospital for the Insane, Washington, District of Columbia.

Dr. R. L. Parsons, New York City Lunatic Asylum, New York.

Dr. Mark Ranney, Iowa Hospital for the Insane, Mt. Pleasant, Iowa.

Dr. Isaac Ray, Philadelphia, Pennsylvania.

Dr. Joseph A. Reed, Western Pennsylvania Hospital for the Insane, Dixmont, Allegheny county, Pennsylvania.

Dr. Henry Riedel, Ward's Island Emigrant Hospital for the Insane, New York.

Dr. F. P. Roy, Quebec Lunatic Asylum.

Dr. John W. Sawyer, Butler Hospital, Providence, Rhode Island.

Dr. A. M. Shew, General Hospital for the Insane, Middletown, Connecticut.

Dr. John Waddell, Provincial Lunatic Asylum, St. John, New Brunswick.

Dr. Clement A. Walker, Lunatic Hospital, Boston, Massachusetts.

Dr. John W. Whitney, Eastern Lunatic Asylum, Lexington, Kentucky.

Dr. Joseph Workman, Asylum for the Insane, Toronto, Canada.

Also the following gentlemen by invitation :

Dr. George Brown, Private Institution for Feeble Minded Youth, Barre, Massachusetts.

Samuel D. Hastings, Secretary State Board of Charities and Reform, Wisconsin.

Dr. Edward R. Hun, Albany, New York.

J. W. Langmuir, Esq., Inspector of Asylums, &c., Province of Ontario.

H. M. Skillman, Commissioner of the Eastern Lunatic Asylum, Lexington, Kentucky.

Dr. H. B. Wilbur, New York Asylum for Idiots, Syracuse, New York.

Dr. Curwen gave a sketch of the life of Dr. N. D. Benedict, and offered the following resolutions, which were unanimously adopted.

Resolved, That this Association has heard, with deep and unfeigned regret, of the death of our late associate and member, Dr. N. D. Benedict, who was for many years an earnest, faithful, highly esteemed, and greatly beloved member of this Association; and that we most sincerely sympathize with his family in the great loss which they have sustained, and rejoice that they and we have an example in his life of all that was noble, pure and true.

Resolved, That a copy of this resolution be forwarded to the family.

The following resolutions, offered by Dr. Kirkbride, were unanimously adopted.

Resolved, That this Association re-affirms, in the most emphatic manner, its former declarations in regard to the construction and organization of Hospitals for the Insane ; and it would take the present occasion to add that, at no time since these declarations were originally made, has anything been said or done to change, in any respect its frequently expressed and unequivocal conviction on the following points, derived, as they have been, from the patient, varied and long-continued observations of its members :

First. That a very large majority of those suffering from mental disease can no where else be as well or as successfully cared for, for the cure of their maladies, or be made as comfortable, if not curable, with equal protection to the patient and the community, as in well-arranged hospitals, specially provided for the treatment of the insane

Second. That neither humanity, economy or expediency can make it desirable that the care of the recent and chronic insane should be in separate institutions.

Third. That those institutions, especially if provided at the public cost, should always be of a plain but substantial character ; and while characterized by good taste, and furnished with everything essential to the health and comfort and successful treatment of the patients, all extravagant embellishments and every unnecessary expenditure should be carefully avoided.

Fourth. That no expense that is required to provide just as many of these hospitals as may be necessary to give the most enlightened

care to all their insane can properly be regarded as either unwise.
inexpedient or beyond the means of any one of the United States
or British Provinces.

A sketch of the life of Dr. John Fonerden, prepared by Dr.
Stribling, at the request of the Association, was read at this
meeting.

Madison, Wisconsin, was selected as the next place of meeting.
on the last Tuesday of May, 1872.

Dr. Nathan D. Benedict was born in Otsego county, New York,
on April 7, 1815. Graduated with honor in 1837, at Rutgers
College, New Brunswick, New Jersey, and commenced the study
of medicine immediately after.

He graduated at the University of Pennsylvania in the spring
of 1841, and at once engaged in practice in Philadelphia, where
he was successfully pursuing his profession when he was appointed
Medical Superintendent of the Philadelphia Almshouse in 1846.

He was chosen Superintendent of the State Lunatic Asylum at
Utica, New York, in the fall of 1849. While engaged in directing the necessary alterations for the heating and ventilation of
that building, he was taken with pneumonia, attended by profuse
hemorrhage, and when able to be about after many months of
confinement, he was recommended to spend the winter of 1853-4
in Florida. Resigning his position with the greatest reluctance,
for his heart was in the work, he removed to Florida in the fall
of 1855, and opened an Institution for Invalids at Magnolia. In
this he succeeded well until the breaking out of the rebellion,
which virtually compelled him to give up his intentions, as the
Government took charge of his buildings for hospital purposes,
and he removed to St. Augustine, where he continued to reside
and filled several offices of honor and trust.

He died on April 30, 1871.

Dr. John Fonerden was born in the city of Baltimore, in the
year 1802.

He commenced the practice of medicine in that city, and in
the earlier portion of his professional life, devoted himself es-

pecially to midwifery, and became in this line, one of the most popular and most reliable practioners in the city.

He was elected Resident Physician of the Maryland Hospital in June, 1846, and continued in that position until his death in April, 1869, greatly respected and esteemed by all who knew him.

Dr. Kirkbride, from the Committee on Didactic and Clinical Instruction in Insanity, offered the following resolutions which were unanimously adopted :

Resolved, That in view of the frequency of mental disorders among all classes and descriptions of people, and in recognition of the fact that the first care of nearly all of these cases necessarily devolves upon physicians engaged in general practice, and this at a period when sound views of the disease and judicious modes of treatment are specially important, it is the unanimous opinion of this Association that in every school conferring medical degrees, there should be delivered by competent professors, a complete course of lectures on insanity and on medical jurisprudence, as connected with disorders of the mind.

Resolved, That these lectures should be delivered before all the students attending these schools, and that no one should be allowed to graduate without as thorough an examination on these subjects as in the other branches taught in the schools.

Resolved, That in connection with these lectures, whenever practicable, there should be clinical instructions, so arranged that, while giving the student practical illustrations of the different forms of insanity and the effects of treatment, they should in no way be detrimental to the patients.

The members of the Association visited the Asylum for the Insane at London, Ontario, and held the closing meeting of the Association in that Institution.

———o———

The twenty-sixth annual meeting was held at Madison, Wisconsin, commencing at 10 A. M., of May 23, 1872. The following members were present during the sessions :

Dr. J. P. Bancroft, Asylum for the Insane, Concord, New-Hampshire.

Dr. C. K. Bartlett, Hospital for the Insane, St. Peter, Minnesota.

Dr. D. R. Brower, Eastern Lunatic Asylum, Williamsburg, Virginia.

Dr. John S. Butler, Retreat for the Insane, Hartford, Connecticut.

Dr. R. G. Cabell, Jr., Assistant Physician, Central Lunatic Asylum, Richmond, Virginia.

Dr. J. H. Callender, Hospital for the Insane, Nashville, Tennessee.

Dr. T. B. Camden, Hospital for the Insane, Weston, West Virginia.

Dr. H. F. Carriel, Hospital for the Insane, Jacksonville, Illinois.

Dr. John B. Chapin, Williard Asylum for the Insane, Willard, New York.

Dr. William M. Compton, Lunatic Asylum, Jackson, Mississippi.

Dr. John Curwen, Pennsylvania State Lunatic Hospital, Harrisburg, Pennsylvania.

Dr. T. P. Dudley, Jr., Assistant Physician, Eastern Lunatic Asylum, Lexington, Kentucky.

Dr. J. F. Ensor, Lunatic Asylum, Columbia, South Carolina.

Dr. F. T. Fuller, Assistant Physician, Asylum for the Insane, Raleigh, North Carolina.

Dr. John P. Gray, State Lunatic Asylum, Utica, New York.

Dr. William Hamilton, Assistant Physician, Western Lunatic Asylum, Staunton, Virginia.

Dr. W. W. Hester, Aesistant Physician, Hospital for the Insane, Indianapolis, Indiana.

Dr. C. H. Hughes, State Lunatic Asylum, Fulton, Missouri.

Dr. Edward A. Kilbourne, Hospital for the Insane, Elgin, Illinois.

93

Dr. Thomas S. Kirkbride, Pennsylvania Hospital for the Insane, Philadelphia, Pennsylvania.

Dr. Henry Landor, Lunatic Asylum, London, Ontario.

Dr. J. M. Lewis, Northern Ohio Lunatic Asylum, Newburg, Ohio.

Dr. A. S. McDill, Hospital for the Insane, Madison, Wisconsin.

Dr. Charles H. Nichols, Government Hospital for the Insane, Washington, District of Columbia.

Dr. R. J. Patterson, Bellevue Place, Batavia, Illinois.

Dr. William L. Peck, Central Ohio Lunatic Asylum, Columbus, Ohio.

Dr. Mark Ranney, Hospital for the Insane, Mt. Pleasant, Iowa.

Dr. D. D. Richardson, Department for the Insane, Almshouse, Philadelphia, Pennsylvania.

Dr. Henry Riedel, Emigrant Hospital for the Insane, Ward's Island, New York.

Dr. John W. Sawyer, Butler Hospital, Providence, Rhode Island.

Dr. A. M. Shew, General Hospital for the Insane, Middletown, Connecticut.

Dr. G. A. Shurtleff, Asylum for the Insane, Stockton, California.

Dr. Charles W. Stevens, County Lunatic Asylum, St. Louis, Missouri.

Dr. William. F. Steuart, Maryland Hospital, Baltimore, Maryland.

Dr. E. H. Van Deusen, Asylum for the Insane, Kalamazoo, Michigan.

Dr. C. A. Walker, Lunatic Hospital, Boston, Massachusetts.

Dr. Joseph J. Webb, Longview Asylum; Cincinnati, Ohio.

Dr. James W. Wilkie, State Lunatic Asylum for Insane Criminals, Auburn, New York.

Dr. Joseph Workman, Asylum for the Insane, Toronto, Ontario.

Dr. Joshua H. Worthington, Friends' Asylum, Frankford, Philadelphia, Pennsylvania.

94

Dr. J. H. Woodburn, Indianapolis, Indiana.

Also by invitation :

Dr. Genet Conger, Trustee of Willard Asylum for the Insane. Willard, New York.

Rev. A. H. Kerr, of St. Peter, Minnesota, Secretary of the Board of Trustees of the State Hospital for the Insane of Minnesota.

Mr. M. L. Fisher, President of the Board of Trustees of the Iowa State Hospital for the Insane.

Dr. Brown, of Madison, Wisconsin.

W. R. Taylor, E. W. Young, Trustees of the Wisconsin State Hospital for the Insane.

Dr. John Faville, President of the State Medical Society of Wisconsin.

Dr. William M. Compton gave a biographical sketch of Dr. A. B. Cabaniss.

Alfred B. Cabaniss was born in the city of Huntsville, in the State of Alabama, on the 7th day of December, 1808, and died in Hinds county, Mississippi, on the 21st day of November, 1871, not quite sixty-three years old. Dr. Cabaniss received a diploma from the Transylvania University at Lexington, Kentucky, in 1833, and in 1835 was admitted to the ad eundem degree at the Jefferson College in Philadelphia. He settled in the town of Raymond, in Hinds county, more than thirty years ago, and afterwards removed to the city of Jackson, where he made a reputation for skill and kindness not surpassed by any member of the profession in Mississippi.

During the war he was not an idle spectator, but at an early day offered his services to the sick and wounded Confederate soldiers, and for a long time was Post Surgeon at the city of Jackson.

Soon after the war he was appointed Superintendent of the Mississippi State Lunatic Asylum. While he was "the good man of the House," about four years, we know that he attached to himself, not only the employes of the household, but the patients also, who regarded him as their father. Nowhere, perhaps, except in his own immediate family circle, did his death cast a sadder gloom than it did upon the household of the Lunatic Asylum.

Dr. Kirkbride presented resolutions in regard to overcrowding hospitals, which were unanimously adopted, as follows :

Resolved, That this Association regards the custom of admitting a greater number of patients than the buildings can properly accommodate, which is now becoming so common in hospitals for the insane in nearly every section of the country, as an evil of great magnitude, productive of extraordinary dangers, subversive of the good order, perfect discipline and greatest usefulness of these institutions, and of the best interests of the insane.

Resolved, That this Association, having repeatedly affirmed its well-matured convictions of the humanity, expediency and economy of every State making ample provision for all its insane, regards it as an important means of effecting this object that these institutions should be kept in the highest state of efficiency, and the difference in condition of patients treated in them and those kept in alms-houses, jails, or even private houses, be thus most clearly demonstrated.

Resolved, That while fully recognizing the great suffering and serious loss that must result to individuals by their exclusion from hospitals when laboring under an attack of insanity, this Association fully believes that the greatest good will result to the greatest number, and at the earliest day, by the adoption of the course now indicated.

Resolved, That the boards of management of the different hospitals on this continent, be urged, most earnestly, to adopt such measures as will effectually prevent more patients being admitted into their respective institutions, than, in the opinions of their superintendents, can be treated with the greatest efficiency, and without impairing the welfare of their fellow sufferers.

Resolved, That the Secretary be instructed to furnish a copy of these resolutions to the boards of management of the different hospitals for the insane in the United States and British Provinces.

A committee was appointed to report on the subject of a competent allowance to the officers of institutions for the insane who have served a term of years in their respective institutions and, when partially incapacitated, are compelled to resign.

Baltimore, Maryland, was selected as the place of next meeting, on the fourth Tuesday of May, 1873.

The Association visited the Hospital for the Insane, the University of Wisconsin, and other objects of interest in and around Madison.

————o————

The twenty-seventh annual meeting of the Association was held in the city of Baltimore, Maryland, commencing at 10 A. M., of May 27, 1873. The following members were present during the session :

Dr. J. P. Bancroft, Asylum for the Insane, Concord, New Hampshire.

Dr. J. W. Barstow, Sanford Hall, Flushing, New York.

Dr. C. K. Bartlett, Hospital for the Insane, St. Peter, Minnesota.

Dr. D. R. Brower, Eastern Lunatic Asylum, Williamsburg, Virginia.

Dr. D. Tilden Brown, Bloomingdale Asylum, New York City.

Dr. George Syng Bryant, First Kentucky Lunatic Asylum, Lexington, Kentucky.

Dr. John S. Butler, Hartford, Connecticut.

Dr. John H. Callender, Hospital for the Insane, Nashville, Tennessee.

Dr. Edward R. Chapin, Kings County Lunatic Asylum, Flatbush, New York.

Dr. John B. Chapin, Willard Asylum, Willard, New York.

Dr. William M. Compton, State Lunatic Asylum, Jackson, Mississippi.

Dr. D. B. Conrad, Central Lunatic Asylum, Richmond, Virginia.

Dr. John Curwen, Pennsylvania State Lunatic Hospital, Harrisburg, Pennsylvania.

Dr. James H. Denny, Retreat for the Insane, Hartford, Connecticut.

Dr. William H. DeWitt, Assistant Physician, Longview Asylum, Carthage, Ohio.

Dr. Joseph Draper, Asylum for the Insane, Brattleboro', Vermont.

Dr. B. D. Eastman, Lunatic Hospital, Worcester, Massachusetts.

Dr. Pliny Earle, Lunatic Hospital, Northampton, Massachusetts.

Dr. M. G. Echeverria, New York.

Dr. Orpheus Everts, Hospital for the Insane, Indianapolis, Indiana.

Dr. F. T. Fuller, Assistant Physician, Insane Asylum, Raleigh, North Carolina.

Dr. John P. Gray, State Lunatic Asylum, Utica, New York.

Dr. Thomas F. Greene, State Lunatic Asylum, Milledgeville, Georgia.

Dr. Richard Gundry, Lunatic Asylum, Athens, Ohio.

Dr. William B. Hazard, Lunatic Asylum, St. Louis, Missouri.

Dr. James C. Hallock, State Emigrant Hospital for the Insane, Ward's Island, New York.

Dr. George F. Jelly, McLean Asylum, Somerville, Massachusetts.

Dr. Thomas S. Kirkbride, Pennsylvania Hospital for the Insane, Philadelphia, Pennsylvania.

Dr. A. H. Kunst, Assistant Physician, Hospital for Insane, Weston, West Virginia.

Dr. Henry Landor, Asylum for the Insane, London, Ontario.

Dr. Edward Mead, Psychopathic Retreat, Winchester, Massachusetts.

Dr. S. J. T. Miller, Southern Ohio Lunatic Asylum, Dayton, Ohio.

Dr. Charles H. Nichols, Government Hospital for the Insane, Washington, District of Columbia.

Dr. Isaac Ray, Philadelphia, Pennsylvania.

Dr. Joseph A. Reed, Western Pennsylvania Hospital for the Insane, Dixmont, Allegheny County, Pennsylvania.

Dr. F. E. Roy, Lunatic Asylum, Quebec, Canada.

Dr. John W. Sawyer, Butler Hospital, Providence, Rhode Island.

Dr. S. S. Schultz, State Hospital for the Insane, Danville, Pennsylvania.

Dr. A. M. Shew, General Hospital for the Insane, Middletown, Connecticut.

Dr. G. A. Shurtleff, Insane Asylum, Stockton, California.

Dr. T. R. H. Smith, State Lunatic Asylum, Fulton, Missouri.

Dr. R. S. Steuart.

Dr. William J. Steuart, Maryland Hospital, Catonsville, Maryland.

Dr. William H. Stokes, Mount Hope Retreat, Baltimore, Maryland.

Dr. Francis T. Stribling, Western Lunatic Asylum, Staunton, Virginia.

Dr. J. D. Thomson, Mt. Hope Retreat, Baltimore, Maryland.

Dr. John E. Tyler, Boston, Massachusetts.

Dr. C. A. Walker, Lunatic Hospital, Boston, Massachusetts.

Dr. E. H. Van Deusen, Asylum for the Insane, Kalamazoo, Michigan.

Dr. R. M. Wigginton, Assistant Physician, Hospital for the Insane, Madison, Wisconsin.

Dr. James W. Wilkie, State Lunatic Asylum for Insane Criminals, Auburn, New York.

Dr. J. H. Worthington, Friends' Asylum for the Insane, Frankford, Philadelphia, Pennsylvania.

Also by invitation, Dr. J. J. Mayeda, of Japan ; Dr. A. S. Ashmead, of Philadelphia.

Dr. Butler resigned the office of President, and the following officers were elected :

Dr. Charles H. Nichols, President.

Dr. C. A. Walker, Vice President.

Dr. Ray read a paper descriptive of the qualifications of officers of Hospitals for the Insane, which was ordered to be printed at the expense of the Association.

The following resolutions were unanimously adopted :

WHEREAS, The President of the Board of Charities of Pennsylvania has requested that this Association should express its opinion in regard to the proper disposition of insane convicts : therefore,

Resolved, 1. That neither the cells of penitentiaries and jails, nor the wards of ordinary hospitals for the insane are proper places for the custody and treatment of this class of the insane.

2. That when the number of this class in any State (or in any two or more adjoining States which will unite in the project) is sufficient to justify such a course, these cases should be placed in a hospital specially provided for the purpose ; and that until this can be done, they should be treated in a hospital connected with some prison, and not in the wards or in separate buildings upon any part of the grounds of an ordinary hospital for the insane.

Nashville, Tennessee, was selected as the next place of meeting, on the third Tuesday of May, 1874.

The Association visited the Maryland Hospital, Mt. Hope Retreat, and the Shephard Hospital for the Insane.

———o———

The twenty-eighth annual meeting was held in the city of Nashville, Tennessee, commencing at 10 A. M., of Tuesday, May 19, 1874. The following members were present during the sessions of the Association :

Dr. Judson B. Andrews, Assistant Physician, State Lunatic Asylum, Utica, New York.

Dr. H. M. Bassett, Iowa Hospital for the Insane, Mt. Pleasant, Iowa.

Dr. J. E. Bowers, Assistant Physician, Hospital for the Insane, St. Peter, Minnesota.

Dr. George Syng Bryant, First Kentucky Lunatic Asylum, Lexington, Kentucky.

Dr. R. G. Cabell, Jr., Assistant Physician, Central Lunatic Asylum, Richmond, Virginia.

Dr. John H. Callender, Hospital for the Insane, Nashville, Tennessee.

Dr. T. B. Camden, West Virginia Hospital for the Insane, Weston, West Virginia.

Dr. H. F. Carriel, State Hospital for the Insane, Jacksonville, Illinois.

Dr. William M. Compton, State Lunatic Asylum, Jackson, Mississippi.

Dr. John Curwen, Pennsylvania State Lunatic Hospital, Harrisburg, Pennsylvania.

Dr. B. D. Eastman, Worcester Lunatic Hospital, Worcester, Massachusetts.

Dr. Orpheus Everts, Indiana Hospital for the Insane, Indianapolis, Indiana.

Dr. Edward C. Fisher, Assistant Physician, Western Lunatic Asylum, Staunton, Virginia.

Dr. C. C. Forbes, Central Kentucky Lunatic Asylum, Anchorage, Kentucky.

Dr. F. T. Fuller, Assistant Physician Insane Asylum, Raleigh, North Carolina.

Dr. Thomas F. Green, Georgia State Lunatic Asylum, Milledgeville, Georgia.

Dr. Charles H. Hughes, St. Louis, Missouri.

Dr. George F. Jelley, McLean Asylum, Somerville, Massachusetts.

Dr. William P. Jones, Nashville, Tennessee.

Dr. E. A. Kilbourne, Hospital for the Insane, Elgin, Illinois.

Dr. Stephen Lett, Assistant Physician, Asylum for the Insane, London, Ontario, Canada.

Dr. William L. Peck, Cincinnati Sanatarium, College Hill, Hamilton County, Ohio.

Dr. Mark Ranney, Hospital for the Insane, Madison, Wisconsin.

Dr. A. Reynolds, Iowa Hospital for the Insane, Independence, Iowa.

Dr. James Rodman, Second Kentucky Lunatic Asylum, Hopkinsville, Kentucky.

Dr. Abram Marvin Shew, General Hospital for the Insane, Middletown, Connecticut.

Dr. Lewis Slusser, Northern Ohio Hospital for the Insane, Newburg, Ohio.

Dr. T. R. H. Smith, State Lunatic Asylum, No. 1, Fulton, Missouri.

Dr. Charles F. Stewart, Nebraska Hospital for the Insane, Lincoln, Nebraska.

Dr. Charles W. Stevens, St. Louis, Missouri.

Dr. Clement A. Walker, Boston Lunatic Hospital, Boston, Massachusetts.

Dr. D. R. Wallace, Texas Lunatic Asylum, Austin, Texas.

Dr. J. F. Webb, Longview Asylum, Carthage, Ohio.

Dr. James W. Wilkie, State Lunatic Asylum for Insane Criminals, Auburn, New York.

Dr. J. H. Worthington, Friends' Asylum for the Insane, Philadelphia, Pennsylvania.

Dr. J. A. Wallis, of the Lunatic Asylum of the County of Durham, England ; Dr. Boyd, of the new Hospital for the Insane, of East Tennessee, and Dr. Brannoch, of the new Hospital for the Insane of West Tennessee, were invited to participate in the deliberations of the Association.

A biographical sketch of Dr. William H. Rockwell, prepared by Dr. Joseph Draper, was read.

"Dr. William H. Rockwell, late Superintendent of the Vermont Asylum for the Insane, died at that institution on the 30th day of November, 1873, after a protracted illness of eighteen months. He was, at the time of his resignation, in August, 1872, the oldest superintendent of an asylum in the United States, having received his appointment on the 28th of June, 1836, and been in active service for more than thirty-six years.

He was a native of East Windsor, Connecticut, and born February 15, 1800. He graduated at Yale College in 1824, and at the medical department of the same institution in 1831.

In 1827, and while a student of Dr. Hubbard, of Pomfret. Connecticut, he received the appointment of assistant to Dr. Todd, at the Hartford Retreat. He remained connected with the Retreat most of the time until his appointment to the Vermont Asylum.

During the illness of Dr. Todd, and after his death, he had charge of the Retreat until the appointment of Dr. Fuller, and wrote the report for the year 1834.

He was married June 25, 1835, to Mrs. Maria J. Chapin, of Coventry, Connecticut. They had three children, a daughter and two sons, the youngest, Captain Charles J. Rockwell, graduated at West Point in 1863, and died in Washington, District of Columbia, of typho marlarial fever in 1867. The elder, Dr. W. H. Rockwell, Jr., was associated with his father in the asylum, as his assistant, for ten years, and was elected his successor. in 1872, but resigned that position at the end of the year.

Dr. Rockwell was an early member of the Association. He was prevented from attending the organization by reason of a bill at that time pending legislation in his own State which exacted his attention in behalf of the interests of the insane who might be committed to his care.

He wrote but little ; his annual reports were noted for brevity, and confined mainly to the results of each year. His was a practical life, he indulged little in theorizing, but was wont to detail his actual observations, and to those associated with him he gave freely of his professional and practical experience.

Few men possessed such qualifications for surmounting difficulties as he, and the history of the Institution at Brattleboro' gives tangible evidence of his indefatigable energy. He was pre-eminently self-reliant, and though he differed from some of his colleagues in the matter of policy in practical management. he was scrupulously faithful to his convictions and to his trusts.

For a year and a half preceding his death he was confined to his bed, suffering most from his fractured limb, gradually wearing away, and sinking to his final rest ; and then it was that the strong points of his character shone out with the most striking

brilliancy. Realizing that his work was done, and that he had
done it faithfully, he expressed his willingness to be judged by
it ; undisturbed by the shafts of malice and indiscriminate censure, he calmly observed : " That his work would be better appreciated, and his motives better understood, after he had gone."
And so he passed away ; dying as he had lived, strong in the
faith of his life-long convictions, and relying with unshaken confidence upon the Divine justice which metes out to every man the
full measure of his deserts.

The following resolutions, prepared by Dr. Green, were unanimously adopted :

WHEREAS, The Association of Medical Superintendents of American Institutions for the Insane has received information of the
death, since their last meeting, of Dr. William H. Rockwell, of Brattleboro', Vermont, who, for thirty-six years has been an earnest,
faithful and efficient laborer in the noblest field of benevolence connected with the healing art.

Be it resolved, That in the death of Dr. Rockwell, the interests of
suffering humanity, in its most fearful form, have sustained a loss
greatly to be deplored, and this Association an able co-worker, counsellor and friend.

Resolved further, That to the family and friends of the deceased we
would tenderly offer our condolence and sympathy in this their sad
bereavement.

Resolved, That the Secretary be requested to furnish to the family
of Dr. Rockwell a copy of this testimonial of our appreciation of the
deceased and profound regret at his death.

The following biographical sketch of Dr. Charles E. Van Anden prepared by Dr. James W. Wilkie, was also read :

" Dr. Charles E. Van Anden was born in Auburn, New York,
January 9, 1819, and, with a few brief absences, spent his whole
life there. He was the son of one of the earliest settlers and
most respected citizens of Auburn. He entered Union College
in 1835, and held during his entire college course, a highly respectable position in his class, graduating August 9, 1839. He
there laid the foundation of those refined and scholarly tastes
which characterized his later years, and which were so well known
and appreciated by his more intimate friends. After leaving col-

lege. he spent some time as a private tutor in the city of New York, and later as a student of theology, with the late Dr. Croswell, then of Auburn. As a student of theology, he won the love and esteem of that distinguished and warm-hearted divine. For reasons quite satisfactory to himself, he gave up the study of theology, and became a student of medicine in the office of Dr. Lansing Briggs, of Auburn, and received the degree of Doctor of Medicine at the Buffalo University, in 1850, having previously attended two courses of lectures at the Geneva Medical College.

He then opened an office for the practice of his profession in Auburn, and early attracted the attention of Dr. Joseph T. Pitney, then in extensive practice as a surgeon, and won from him his highest esteem professionally, as well as his warmest personal regards. Dr. Pitney's love and appreciation of him continued through life.

In 1852, Dr. Van Anden was called to take charge of the Cholera Hospital at Buffalo, at a time when that terrible malady was making great havoc in that city. After consulting with his friends in Auburn, he came to the conclusion that it was a call of duty, and unhesitatingly entered into the midst of the pestilence, and by his calm and dignified Christian deportment, and the wise exercise of his skill as a physician, won the esteem and approbation of all with whom he came in contact.

In 1857, he was appointed Physician to the Auburn State Prison, and in 1859 was appointed assistant to Dr. Edward Hall, then Superintendent of the State Lunatic Asylum for Insane Convicts at Auburn, and, on Dr. Hall's retirement, in 1862, succeeded to that responsible position. This position he held until 1870, eight years.

Since that time Dr. Van Anden devoted his time to the practice of his profession in Auburn. Modest, sensitive and distrustful of his own abilities, he lacked that energy of purpose and those aggressive qualities so requisite to success. Hence the self advertised quack, pushing his own claims, was quite likely to outstrip him in the race for popular favor. But a work placed in his hands was performed with the greatest intelligence and fidelity.

In his manners he was dignified, but courteous, his affability and kindness winning the hearts of those with whom he was most intimate. In general knowledge, in sound judgment, in all the graces of refinement and scholarly cultivation, Dr. Van Anden excelled. In private life, of the greatest purity of character, he maintained a spotless reputation as a public officer. He died a poor, but honest, man.

At the time of his death Dr. Van Anden was a member of the New York State Medical Society, and of the Medical Society of Cayuga County, in which he lived.

His death occurred October 19, 1873, and was the result of a peculiar and distressing accident. Eight days previously, as he was about to retire for the night, he unconsciously drew into the œsophagus a rubber plate, of triangular form, about an inch in diameter, to which was attached a single false tooth. After making several unsuccessful attempts to remove it with the œsophagus forceps, he applied to his former preceptor, Dr. Briggs, who also failed to detect its location with the forceps, and remove it. The next morning he introduced a probang, and supposed he had dislodged and pushed it forward into the stomach. Violent inflammation supervened, with swelling and inability to swallow. Dr. E. M. Moore, of Rochester, visited the patient on the fourth day, when the inflammation and swelling were so great that he deemed an exploration of the œsophagus impractible. His strength was sustained by injections of beef tea, &c., until the eighth day, when profuse hemorrhage took place, from which he sank and died.

An autopsy revealed the plate concealed just within the œsophagus, a sharp angle of which had made an incision one-half an inch in length through its posterior wall. Near the base of the lung was a gangrenous mass, involving to a considerable extent, the tissues of the lung itself, and which was the seat of hemorrhage. In attempting to swallow, liquid aliment was forced through the aperture in the œsophagus, which infiltrated itself through the cellular tissues, and gravitating to the point mentioned, had excited inflammation that resulted in gangrene and death.''

The President appointed Dr. Shew committee to prepare resolutions expressive of the sense of the Association ; who subsequently offered the following resolutions :

Resolved, That the Association has received the announcement of the death of Dr. Charles E. Van Anden, formerly Superintendent of the Asylum for Insane Criminals at Auburn, New York, and for many years a distinguished member of this Association, with the deepest concern.

Resolved, That apart from high professional reputation always enjoyed by Dr. Van Anden, both as a practitioner of medicine and Superintendent of the asylum, his excellent private character, his many Christian virtues, his uniform courtesy and honorable intercourse with his fellows, have endeared him to the members of this Association, as well as to a large circle of admiring friends.

Resolved, That while the Association deeply sympathize with his family in their bereavement, they, with all his other friends, feel confident that when time has softened the sadness of parting, the memory of his life will be an enduring source of comfort and pleasure to those from whom he has been taken away.

Resolved, That a copy of these resolutions be presented to the family of our late associate, signed by the officers of the Association.

The following resolution, offered by Dr. William P. Jones, was unanimously adopted :

WHEREAS, It has formally been brought to the notice of the Association that State and county authorities, having supreme direction of institutions for the insane, have, by law, departed from the spirit and tenor of the principles and general regulations for their government which, after observation, experience and mature deliberation, have been promulgated and recommended by this body as judicious and humane ; therefore,

Resolved, That we reaffirm former utterances of the Association, as fully expressive of our views as to the proper manner of conducting hospitals for the insane, and that we earnestly commend these utterances to the favorable consideration and regard of the managers of asylums throughout the country.

The greater part of the time of the Association was taken up in hearing reports from members of the progress made in the care and management of the insane and in discussing the most recent modes of medical treatment.

The Association adjourned to meet in Stockton, California, on the third Tuesday in May, 1875.

Dr. Francis T. Stribling was born January 20, 1810, in the town of Staunton, Virginia, where he received his elementary education, and soon entered the office of his father, who was then clerk of the county of Augusta, in which he remained several years. It was probably in this position, that he acquired those habits of neatness, method, and order for which he was so much distinguished in after life. Having determined to adopt the medical profession, after some preparatory reading under the advice of a distinguished physician of Staunton, he spent a session at the University of Virginia, and in the following year took his degree in Philadelphia. He then commenced the practice of his profession in Staunton, and soon won the confidence of the public.

In 1836, at the early age of twenty-six years, he was elected by the distinguished gentlemen, who then composed the Board of Directors, Physician to the Western Lunatic Asylum. Within a few days after his election, Dr. Stribling went on a tour of observation through the middle and northern States, to inspect the best regulated institutions for the insane, and to gather by observation and intercourse with those in charge, all the information necessary to guide him in the discharge of his responsible duties.

He returned from that tour with expanded views, and much valuable information, and became indoctrinated with the views of those who had materially advanced opinions, as to the curability of insanity in a large proportion of cases where the disease proceeded from functional causes, and returned home an enthusiast in the great work, to which his future life was to be dedicated.

In the autumn of 1836, he made known to the Board of Directors, his views of the proper management of the institution, and invoked their assistance in appealing to the Legislature for the means of carrying them out. The Board promptly and cheerfully acceded to the request, and an appeal made to the Legislature in December by the Board of Directors, was generously responded to by that body, at that and subsequent sessions. In his annual reports of a later date, Dr. Stribling pointed out many imperfections in the laws regulating the Asylums for the Insane.

During the sessions of 1840-41, the Committee of the Legislature to which his recommendation had been referred, feeling their incompetency to deal intelligently with the subject, summoned the Physicians of the two Asylums, and a member of the Board of Directors of each, to appear before the Committee, to give their assistance in making a thorough revision of all the laws relating to the asylums. Upon receiving this summons, Dr. Stribling, with the aid of the members of the Board designated to accompany him, before leaving home, prepared a bill which covered the whole subject.

This bill was presented to the Committee and, without material amendment, was reported and soon afterwards received the sanction of both houses, and still remains in force. Dr. Stribling may therefore be considered the author of that law.

From that time to the commencement of his last illness, he devoted himself assiduously to the enlargment of the capacities of the institution for usefulness.

To talents of high order he united unblemished integrity. and warm and generous feelings, while in the discharge of his responsible duties, he exhibited inflexible firmness, and such grace and serenity of manner, as to win the confidence and affection of all who were brought into association with him.

————o————

The Association, at its meeting in Nashville, decided to meet in Stockton, California, on the third Tuesday of May, 1875, but a very large proportion of the members, in the Spring of 1875, decided that they would not go to California, and were in favor of a meeting at Auburn, New York, and in consequence of that decision the twenty-ninth annual meeting was held in Auburn, and was called to order at 10 A. M., of May 18, 1875. After a short address by the President, Dr. Charles H. Nichols, prayer was offered by the Rev. Dr. Condit, of Auburn, New York.

On motion of Dr. Gray, it was resolved that the action of the the officers of the Association, in regard to the change of place of the meeting. be approved.

On motion of Dr. Kirkbride, it was unanimously resolved that Dr. John Charles Bucknill. of England, be elected an honorary member of the Association, and be invited to take a seat and participate in the deliberations of the Association.

Dr. Bucknill expressed his pleasure and gratification at this mark of esteem of the Association.

Hon. Mr. Pomeroy. Mayor of the city of Auburn, then welcomed the Association to the city, to which the President of the Association briefly responded.

The following members were present during the sessions of the Association :

Dr. R. F. Baldwin, Western Lunatic Asylum, Staunton, Virginia.

Dr. Randolph Barksdale, Central Lunatic Asylum, Richmond, Virginia.

Dr. A. T. Barnes, Southern Illinois Hospital for the Insane, Anna, Illinois.

Dr. C. K. Bartlett, Hospital for the Insane, St. Peter, Minnesota.

Dr. J. W. Barstow, Sanford Hall, Flushing, New York.

Dr. D. T. Boughton, Assistant Physician, State Hospital for the Insane, Madison, Wisconsin.

Dr. D. Tilden Brown, Bloomingdale Asylum, New York City.

Dr. W. H. Bunker, Longview Asylum, Carthage, Ohio.

Dr. John H. Callender, Hospital for the Insane, Nashville, Tennessee.

Dr. T. B. Camden, West Virginia Hospital for the Insane, Weston, West Virginia.

Dr. John B. Chapin, Willard Asylum for the Insane, Willard, New York.

Dr. John H. Clark, Western Ohio Hospital for the Insane, Dayton, Ohio.

Dr. William M. Compton, Mississippi State Lunatic Asylum, Jackson, Mississippi.

Dr. J. S. Conrad, Maryland Hospital, Catonsville, Maryland.

Dr. George Cook, Brigham Hall, Canandaigua, New York.

Dr. John Curwen, Pennsylvania State Lunatic Hospital, Harrisburg, Pennsylvania.

Dr. Joseph Draper, Vermont Asylum for the Insane, Brattleboro', Vermont.

Dr. B. D. Eastman, Worcester Lunatic Hospital, Worcester, Massachusetts.

Dr. Orpheus Everts, Indiana Hospital for the Insane, Indianapolis, Indiana.

Dr. F. G. Fuller, State Hospital for the Insane, Lincoln, Nebraska.

Dr. W. W. Godding, Taunton Lunatic Hospital, Taunton, Massachusetts.

Dr. John P. Gray, State Lunatic Asylum, Utica, New York.

Dr. Eugene Grissom, Insane Asylum of North Carolina, Raleigh, North Carolina.

Dr. Richard Gundry, Southeastern Ohio Hospital for the Insane, Athens, Ohio.

Dr. C. H. Hughes, St. Louis, Missouri.

Dr. George F. Jelly, McLean Asylum, Somerville, Massachusetts.

Dr. Thomas S. Kirkbride, Pennsylvania Hospital for the Insane, Philadelphia, Pennsylvania.

Dr. A. H. Knapp, State Insane Asylum, Osawatomie, Kansas,

Dr. Henry Lander, Asylum for the Insane, London, Ontario.

Dr. Joseph D. Lomax, Marshall Infirmary, Troy, New York.

Dr. A. E. Macdonald, New York City Asylum, Ward's Island.

Dr. Carlos F. MacDonald, Brooklyn, E. D. New York.

Dr. Charles H. Nichols, Government Hospital for the Insane, Washington, District of Columbia.

Dr. George C. Palmer, Assistant Physician,Asylum for the Insane, Kalamazoo, Michigan.

Dr. R. L. Parsons, New York City Lunatic Asylum, Blackwell's Island.

Dr. Mark Ranney, Hospital for the Insane, Mt. Pleasant, Iowa.

Dr. Joseph A. Reed, Western Pennsylvania Hospital for the Insane, Dixmont, Pennsylvania.

Dr. A. Reynolds, Hospital for the Insane, Independence, Iowa.

Dr. John W. Sawyer, Butler Hospital, Providence, Rhode Island.

Dr. S. S. Schultz, State Hospital for the Insane, Danville, Pennsylvania.

Dr. Lewis Slusser, Northern Ohio Hospital for the Insane, Cleveland, Ohio.

Dr. Henry R. Stiles, State Homœopathic Asylum for the Insane, Middletown, New York.

Dr. T. R. H. Smith, State Lunatic Asylum No. 1, Fulton, Missouri.

Dr. Henry P. Stearns, Retreat for the Insane, Hartford, Connecticut.

Dr. William H. Stokes, Mount Hope Retreat, Baltimore, Maryland.

Dr. John Waddell, Provincial Lunatic Asylum, St. John, New Brunswick.

Dr. Clement A. Walker, Boston Lunatic Hospital, Boston, Massachusetts.

Dr. D. R. Wallace, State Lunatic Asylum, Austin, Texas.

Dr. James W. Wilkie, State Lunatic Asylum for Insane Criminals, Auburn, New York.

Also the following gentlemen on invitation :

L. Fletcher, Trustee of Hospital for the Insane, St. Peter, Minnesota.

P. H. Miller, Manager Western Pennsylvania Hospital for the Insane, Dixmont, Pennsylvania.

Dr. Samuel Lilly, Commissioner of the State Lunatic Asylum, Morristown, New Jersey.

Dr. H. B. Wilbur, Asylum for Idiots, Syracuse, New York.

Dr. John Ordronaux. Commissioner in Lunacy of New York.
S. H. Jameson, M. D., James S. Athon, M. D., George F.
Chittenden, M. D., J. T. Richardson, M. D., Commissioners of
Indiana Hospital for the Insane.

Col. T. G. Walton, Capt. C. B. Deusen, Commissioners of
the Insane Asylum, Morganton, North Carolina.

In addition to visiting the State Lunatic Asylum for Insane
Convicts, at Auburn, the members of the Association also visited
the Willard Asylum for the Insane, at Willard, on Seneca Lake,
and the State Lunatic Asylum at Utica, New York.

The following resolutions, prepared by Dr. Isaac Ray, were
adopted:

The Association of Medical Superintendents of American Institutions for the Insane, having been formed for the purpose of promoting the welfare of the insane, regard it as one of their duties to
inquire into and pass judgment upon any scheme, project or change,
offered professedly with this end in view. They would be faithless
to the trust they have assumed, were they to remain in silence
while changes in the management of our hospitals are forced upon
us, calculated to impair their usefulness and inflict a positive injury
upon their inmates. The duty to speak at the present time is all the
greater, in view of the fact that the objects sought for by these new
measures, are sufficiently secured in the existing arrangements, and
the pretended demand for them proceeds from no actual, tangible
grievance, but solely from that prevalent spirit of discontent, which is
ever ready to discover a fancied wrong and clamor for a change in
whatever has stood the test of a little time. Were this dissatisfaction confined to the ordinary methods of discussing evils, real or
fancied, it would furnish no ground of complaint, and we would
cheerfully meet it in the same way. But, without reference to us,
without inquiry of any kind, in fact, it has been thrust upon us in
the shape of legislation, unexceptionally mischievious in its effect on
the true purposes of hospitals for the insane, and thus it is that institutions which should be managed on well-matured, intelligent
principles, their course guided by one animating spirit, taking in all
the circumstances of the situation, are disturbed by an intrusive element, having with them no kind of affiliation, and calculated, in the
nature of things, to destroy that harmony of action which is indispensable to the highest measure of success.

Believing that whatever of progress has been accomplished by our hospitals, may be fairly attributed, in a great measure, to the free and independent action allowed to their officers,whereby they have been enabled, without apprehension of popular fear or favor, to manage their charge in the way commended to them, either by the general voice of the profession, or their own deliberate convictions,we should, for that reason alone, deplore any legislation calculated to substitute for such liberty the suggestions of an outside party, entirely ignorant, it may be, of the working of a hospital, as well as of the movement of the insane mind. If the time shall ever come when the Legislature in its zeal for the public good, shall establish a board of officers to supervise the medical practice of the State, with power to enter every sick man's chamber, to inquire respecting the medicine and diet prescribed, and any other matter connected with his welfare, and report the results of their examination to the constituted authorities, then it may be proper to consider the propriety of extending the same kind of paternal visitation to the hospitals for the insane.

Without arrogating to ourselves any extraordinary wisdom, we believe that the accomplished work of this Association, as well as the character and reputation of its present members, fairly entitles it to a respectful hearing on any matters of legislation affecting the interests of the insane in the establishments devoted to their custody and treatment. We, therefore, offer the following resolutions, in the hope that they will receive from the public all the attention to which the importance of the subject, and the authority of the source from which they come, entitle them :

Resolved, That the government of our hospitals, as at present constituted, whereby a physician, supposed to be eminently qualified by his professional training and his traits of character, both moral and intellectual, is invested with the immediate control of the whole establishment, while a Board of Directors, Trustees or Managers, as they are differently called in different places,—men of acknowledged integrity and intelligence—has the general supervision of its affairs, has been found, by ample experience, to furnish the best security against abuses, and the strongest incentives to constant effort and improvement.

Resolved, That any supernumerary functionaries, endowed with the privilege of scrutinizing the management of the hospital, and sitting in judgment on the conduct of attendants and the complaints of patients, and controlling the management, directly by the exercise of superior power, or indirectly by stringent advice, can scarcely accomplish an amount of good sufficient to compensate for the harm that is sure to follow.

Resolved, That the duty of restoring the insane, and of procuring the highest possible degree of comfort for those who are beyond the reach of cure, implies a knowledge of their malady, and of their ways and manners, that can be obtained only by study and observation.

Resolved, That the work of conducting any particular individual through the mazes of disease, into the light of unclouded reason, embracing, as it does, the drugs he is to take, the privileges he is to enjoy, the letters he is to write or receive, and the company he may see, implies not only certain professional attainments, but a close and continuous observation of his conduct and conversation, neither of which qualifications can be expected from the class of functionaries above mentioned, though appointed for the express purpose of making suggestions and proffering advice.

Resolved, That one of the first things in the treatment of a patient is to secure his confidence, to make him feel that he is in the hands of friends who will protect and care for him; and yet this purpose is completely frustrated when it is incessantly proclaimed to him from the walls of his apartment, that the people to whom he has been entrusted, are not trusted by others, and that any aid or comfort he may require, must be sought from a power paramount to theirs.

Resolved, That valuable information may be obtained from the letters of patients respecting their mental movements, as many will communicate their thoughts in this manner more unreservedly than in their conversation, which advantage is lost when their letters are forwarded unopened.

Resolved, That inasmuch as the letters of the insane, especially of women, often contain matter, the very thought of which, after recovery, will overwhelm them with mortification and dismay, any law which compels the sending of such letters, is clearly an outrage on common decency and common humanity.

Resolved, That the fact so much asserted at the present, and offered as the main reason for the legislation in question, viz : that sane persons are often falsely imprisoned on the pretense of insanity, is not true, and that we believe that, if ever, it is extremely rare that a single case of false imprisonment, in any hospital in this country, has taken place.

Resolved, That should such cases occur, it would require more knowledge and experience to detect and expose their true character, than any but the officers of the hospital would be likely to possess.

Resolved, That the Project of Law for regulating the relations of the insane, adopted by the unanimous vote of the Association in 1868, prescribes such safe-guards against abuses of every kind, as are best fitted to secure that object, with the least possible amount of inconvenience to parties not immediately concerned.

Resolved, That the practice, now rather common, even among those who write or lecture on the subject, for the instruction of the public, of designating as " Private Asylums," the corporate hospitals of the country, such as the McLean Asylum, at Somerville, the Butler Hospital, at Providence, the Retreat for the Insane, at Hartford, the Bloomingdale Asylum, in New York, the Friend's Asylum, at Frankford, and the Pennsylvania Hospital, in Philadelphia, is calculated to mislead the public mind respecting the true character of such establishments. Founded, as they are, on the gifts and bequests of benevolent persons, conducted by officers paid by a fixed salary, and Directors or managers with no compensation at all, and watched by a system of visitation, unequaled in frequency and thoroughness, by that of any public hospital, they are in no sense of the term, Private Asylums.

The following resolutions, offered by Dr. Nichols, were also adopted :

1. *Resolved,* That in the opinion of the Association of Medical Superintendents of American Institutions for the Insane, it is the duty of each of the United States, and of each of the Provinces of the Dominion, to establish and maintain a State or public institution for the custody and treatment of inebriates, on substantially the same footing, in respect to organization and support, as that upon which the generality of State and Provincial Institutions for the Insane are organized and supported.

2. *Resolved,* That as, in the opinion of this Association, any system of management of institutions for inebriates, under which the duration of the residence of their inmates, and the character of the treatment to which they are subjected is voluntary on their part, must, in most cases, prove entirely futile, if not worse than useless, there should be in every State and Province such positive constitutional provisions and statutory enactments as will, in every case of presumed inebriety, secure a careful inquisition into the question of drunkenness, and fitness for the restraint and treatment of an institution for inebriates, and such a manner and length of restraint as will render total abstinence from alcoholic or other hurtful stimulants, during such treatment, absolutely certain, and present the best prospects of cure or reform, of which each case is susceptible.

3. *Resolved further*, That the treatment in institutions for the insane of dipsomaniacs, or persons whose only obvious mental disorder is the excessive use of alcoholic or other stimulants, and the immediate effects of each excess, is exceedingly prejudicial to the welfare of those inmates for whose benefit such institutions are established and maintained, and should be discontinued just as soon as other separate provision can be made for inebriates.

———o———

The thirtieth annual meeting of the Association was held in Philadelphia, commencing at 10 A. M., of Tuesday, June 13, 1876. The following members were present :

Dr. William M. Awl, Columbus, Ohio.

Dr. R. F. Baldwin, Western Lunatic Asylum, Staunton, Virginia.

Dr. J. W. Barstow, Sanford Hall, Flushing, New York.

Dr. H. Black, Eastern Lunatic Asylum, Williamsburg, Virginia.

Dr. James A. Blanchard, Kings County Lunatic Asylum, Flatbush, New York.

Dr. D. J. Boughton, Hospital for the Insane, Mendota, Wisconsin.

Dr. D. Tilden Brown, Bloomingdale Asylum, Manhattanville, New York.

Dr. Henry W. Buel, Spring Hill Institution, Litchfield, Connecticut.

Dr. W. H. Bunker, Longview Asylum, Carthage, Ohio.

Dr. John S. Butler, Hartford, Connecticut.

Dr. H. A. Buttolph, State Lunatic Asylum, Morristown, New Jersey.

Dr. R. C. Cabell, Jr., Assistant Physician, Central Lunatic Asylum, Richmond, Virginia.

Dr. John H. Callender, Hospital for the Insane, Nashville, Tennessee.

Dr. T. B. Camden, Hospital for the Insane, Weston, West Virginia.

Dr. H. F. Carriel, Hospital for the Insane, Jacksonville, Illinois.

Dr. George C. Catlett, Lunatic Asylum, No. 2, St. Joseph, Missouri.

Dr. John B. Chapin, Willard Asylum for the Insane, Willard, New York.

Dr. R. C. Chenault, Eastern Lunatic Asylum, Lexington, Kentucky.

Dr. W. S. Chipley, Cincinnati Sanitarium, College Hill, Ohio.

Dr. Daniel Clark, Asylum for the Insane, Toronto, Ontario.

Dr. William M. Compton, Lunatic Asylum, Jackson, Mississippi.

Dr. John Curwen, Pennsylvania State Lunatic Hospital, Harrisburg, Pennsylvania.

Dr. James H. Denny, New York.

Dr. J. T. Ensor, Asylum for the Insane, Columbia, South Carolina.

Dr. Orpheus Everts, Hospital for the Insane, Indianapolis, Indiana.

Dr. C. C. Forbes, Central Kentucky Lunatic Asylum, Anchorage, Kentucky.

Dr. F. G. Fuller, Hospital for the Insane, Lincoln, Nebraska.

Dr. John P. Gray, State Lunatic Asylum, Utica, New York.

Dr. Thomas F. Greene, Lunatic Asylum, Milledgeville, Georgia.

Dr. Eugene Grissom, Insane Asylum, Raleigh, North Carolina.

Dr. Richard Gundry, Hospital for the Insane, Athens, Ohio.

Dr. Henry M. Harlow, Hospital for the Insane, Augusta, Maine.

Dr. J. Welch Jones, Lunatic Asylum, Jackson, Louisiana.

Dr. Walter Kempster, Hospital for the Insane, Oshkosh, Wisconsin.

Dr. Edwin A. Kilbourne, Hospital for the Insane, Elgin, Illinois.

Dr. Thomas S. Kirkbride, Pennsylvania Hospital for the Insane, Philadelphia, Pennsylvania.

Dr. L. R. Landfear, Hospital for the Insane, Dayton, Ohio.

Dr. A. E. Macdonald, City Asylum for the Insane, Ward's Island, New York.

Dr. C. F. McDonald, State Lunatic Asylum for Insane Criminals, Auburn, New York.

Dr. Edward Mead, Boston, Massachusetts.

Dr. J. W. Mercer, Assistant Physician, Hospital for the Insane, Anna, Illinois.

Dr. Charles H. Nichols, Government Hospital for the Insane, Washington, District of Columbia.

Dr. R. L. Parsons, City Lunatic Asylum, Blackwell's Island, New York.

Dr. Mark Ranney, Hospital for the Insane, Mt. Pleasant, Iowa.

Dr. Isaac Ray, Philadelphia, Pennsylvania.

Dr. A. Reynolds, Hospital for the Insane, Independence, Iowa.

Dr. D. D. Richardson, Department for the Insane, Almshouse, Philadelphia, Pennsylvania.

Dr. John W. Sawyer, Butler Hospital, Providence, Rhode Island.

Dr. S. S. Schultz, State Hospital for the Insane, Danville, Pennsylvania.

Dr. A. M. Shew, Hospital for the Insane, Middletown, Connecticut.

Dr. T. R. H. Smith, State Lunatic Asylum, No. 1, Fulton, Missouri.

Dr. Henry P. Stearns, Retreat for the Insane, Hartford, Connecticut.

Dr. J. T. Steeves, Provincial Lunatic Asylum, St. John, New Brunswick.

Dr. Clement A. Walker, Boston Lunatic Hospital, Boston, Massachusetts.

Dr. D. R. Wallace, Hospital for the Insane, Austin, Texas.

Dr. J. H. Worthington, Friends' Asylum for the Insane, Philadelphia, Pennsylvania.

The following gentlemen were present by invitation :

Dr. P. O. Hooper, Little Rock, Arkansas.

Mr. D. A. Ogden and Dr. W. A. Swaby, Trustees of the the Willard Asylum, Willard, New York.

Rev. A. H. Kerr and Mr. W. Talbot, Commissioners of the State Hospital for the Insane, St. Peter, Minnesota.

John Sunderland, Superintendent of Construction of the State Hospital for the Insane, Warren, Pennsylvania.

Mr. J. Whetstone, President of the Board of Trustees of the Cincinnati Sanitarium, College Hill, Ohio,

Mr. John W. Chase, Trustee of the Maine Hospital for the Insane, Augusta, Maine.

Dr. D. H. Kitchen, New York.

Hon. J. W. Langmuir, Inspector of Hospitals, New York.

Dr. T. S. Sumner, New York.

Francis Wells, Esq., Commissioner of the Board of Public Charities, of Pennsylvania.

Mr. F. H. Wines, Secretary of the Board of Public Charities of Illinois.

Drs. Jameson, Richardson and Chittenden, Commissioners of the State Hospital for the Insane, Indianapolis, Indiana.

A biographical sketch of Dr. George Syng Bryant, was read by Dr. R. C. Chenault.

Dr. Bryant was born in Old Virginia, in 1825, and died in June, 1875, in full vigor of manhood. He was educated at Hampden Sydney College, at an early age, it is said, with the honors of his class ; studied medicine and graduated from Old Jefferson, in this city, in 1845. Soon after he removed to Mississippi, where he practiced his profession very successfully for about ten years, up to the commencement of the late civil war, when he was appointed a surgeon in the Confederate service, and won for himself distinction in that service. At the close of the war he removed to St. Louis, Missouri, but was induced to leave that place on account of failing health, brought about by exposure during the war. He removed to Lexington, Kentucky, in the spring of 1868, soon made for himself a reputation as a man of more than ordinary ability, became an active and prominent member of the Kentucky State Medical Society, and won the exalted esteem of the profession generally throughout the State.

His enthusiasm for his profession, his admiration for the masters of his science, his studious habits and his contributions to the various medical journals, all marked him as a man of no ordinary cast. As a gynæcologist, he was distinguished in the west, especially as an operator and also as an inventor.

With those with whom he was associated in the management of the Eastern Kentucky Lunatic Asylum, from the highest to the lowest, all continue to speak of his uniform kindness, and his unceasing efforts to make every one around him comfortable and happy.

He will be missed, indeed, from our Association, from the Kentucky State Medical Society, to which he was a contributor, from the profession generally where he lived, and among whom he had many warm admirers, and from society generally. Therefore,

Resolved, That this Association tender their warmest sympathy to his personal friends, and especially his widow, Mrs. Bryant, by whom he is missed more than by all others, and to whom he was so much devoted; and we desire that this memorial and resolution be placed upon our minutes, and that the Secretary be requested to forward Mrs. Bryant a copy of the same.

The following members were appointed delegates to the International Medical Congress, which met in Philadelphia on September 4, 1876:

Drs. Thomas S. Kirkbride, Isaac Ray, John Curwen, C. A. Walker, Pliny Earle, John P. Gray, D. Tilden Brown, H. A. Buttolph, Orpheus Everts, Charles H. Nichols, Walter Kempster, Charles H. Hughes, H. F. Carriel, J. H. Callender, W. S. Chipley, James Rodman, Eugene Grissom, C. K. Bartlett, A. M. Shew, James R. DeWolf.

The thirty-first annual meeting was held in St. Louis, Missouri, commencing at 10 A. M., of May 29, 1874. The following members were present:

Dr. A. T. Barnes, Illinois Southern Hospital for the Insane, Anna, Illinois.

Dr. C. K. Bartlett, Minnesota Hospital for the Insane, St. Peter, Minnesota.

Dr. J. K. Banduy, St. Vincent Asylum, St. Louis, Missouri.

Dr. H. Black, Eastern Lunatic Asylum, Williamsburg, Virginia.

Dr. D. T. Boughton, State Hospital for the Insane, Mendota Wisconsin.

Dr, R. M. Bucke, Asylum for the Insane, London, Ontario.

Dr. W. H. Bunker, Longview Asylum, Carthage, Ohio.

Dr. J. H. Callender, Tennessee Hospital for the Insane, Nashville, Tennessee.

Dr. T. B. Camden, Hospital for the Insane, Weston, West Virginia.

Dr. H. F. Carriel, Central Hospital for the Insane, Jacksonville, Illinois.

Dr. George C. Catlett, Lunatic Asylum No. 2, St. Joseph, Missouri.

Dr. John B. Chapin, Willard Asylum for the Insane, Willard, New York.

Dr. W. S. Chipley, Cincinnati Sanitarium, College Hill, Ohio.

Dr. Daniel Clark, Asylum for the Insane, Toronto, Canada.

Dr. William M. Compton, State Lunatic Asylum, Jackson, Mississippi.

Dr. John Curwen, Pennsylvania State Lunatic Hospital, Harrisburg, Pennsylvania.

Dr. Orpheus Everts, Hospital for the Insane, Indianapolis, Indiana.

Dr. F. G. Fuller, State Hospital for the Insane, Lincoln, Nebraska.

Dr. John P. Gray, State Lunatic Asylum, Utica, New York.

Dr. Eugene Grissom, Insane Asylum of North Carolina, Raleigh, North Carolina.

Dr. Richard Gundry, Columbus Hospital for the Insane, Columbus, Ohio.

Dr. William B. Hazard, St. Louis, Missouri.

Dr. H. K. Hinds, Assistant Physician, Lunatic Asylum No. 1. Fulton, Missouri.

Dr. N. De V. Howard, Lunatic Asylum, St. Louis, Missouri.

Dr. C. H. Hughes, St. Louis, Missouri.

Dr. Walter Kempster, Northern Hospital for the Insane, Winnebago, Wisconsin.

Dr. Thomas. H. Kenan, Assistant Physician, Lunatic Asylum, Milledgeville, Georgia.

Dr. E. A. Kilbourne, Northern Hospital for the Insane, Elgin. Illinois.

Dr. L. R. Landfear, Hospital for the Insane, Dayton, Ohio.

Dr. C. F. MacDonald, State Asylum for Insane Criminals, Auburn, New York.

Dr. Andrew McFarland, Oak Lawn Retreat, Jacksonville, Illinois.

Dr. Charles H. Nichols, Government Hospital for the Insane, Washington, District of Columbia.

Dr. Joseph A. Reed, Western Pennsylvania Hospital for the Insane, Dixmont, Pennsylvania.

Dr. James Rodman, Western Kentucky Lunatic Asylum, Hopkinsville, Kentucky.

Dr. John W. Sawyer, Butler Hospital, Providence, Rhode Island.

Dr. Charles W. Stearns, St. Louis, Missouri.

Dr. J. Strong, Cleveland Hospital for the Insane, Newburgh, Ohio.

Dr. Clement A. Walker, Lunatic Hospital, Boston, Massachusetts.

Dr. D. R. Wallace, Asylum for the Insane, Austin, Texas.

Dr. J. M. Wallace, Asylum for the Insane, Hamilton, Ontario.

The following gentlemen were present by invitation :

Mr. S. R. Welles, Trustee of the Willard Asylum, Willard, New York.

Dr. G. F. Chittenden, Commissioner of the Hospital for the Insane, Indianapolis, Indiana.

J. H. Wines, Secretary of Board of Public Charities of Illinois.

Dr. William Corson and Gen. James A. Beaver, Commissioners of the State Hospital for the Insane, Warren, Pa.

Dr. C. F. Wilbur, School for Feeble Minded Children, Jacksonville, Illinois.

———o———

The thirty-second annual meeting was held in Washington, District of Columbia, commencing at 10 A. M., of May 14, 1878. The following members were present.

Dr. R. F. Baldwin, Western Lunatic Asylum, Staunton, Virginia.

Dr. A. T. Barnes, Southern Hospital for the Insane, Anna, Illinois.

Dr. H. Black, Eastern Lunatic Asylum, Williamsburg, Virginia.

Dr. D. T. Boughton, State Hospital for the Insane, Mendota, Wisconsin.

Dr. R. M. Bucke, Asylum for the Insane, London, Ontario.

Dr. D. R. Burrell, Brigham Hall, Canandaigua, New York.

Dr. A. P. Busey, Assistant Physician, Lunatic Asylum, No. 2, St. Joseph, Missouri.

Dr. John H. Callender, Hospital for the Insane, Nashville, Tennessee.

Dr. T. B. Camden, Hospital for the Insane, Weston, West Virginia.

Dr. John B. Chapin, Willard Asylum, Willard, New York.

Dr. W. A. Cheatham, Nashville, Tennessee.

Dr. R. C. Chenault, Eastern Lunatic Asylum, Lexington, Kentucky.

Dr. W. S. Chipley, Cincinnati Sanitarium, College Hill, Ohio.

Dr. Daniel Clark, Asylum for the Insane, Toronto, Ontario.

Dr. William M. Compton, Holly Springs, Mississippi.

Dr. J. S. Conrad, Baltimore, Maryland.

Dr. John Curwen, Pennsylvania State Lunatic Hospital, Harrisburg, Pennsylvania.

Dr. Joseph Draper, Asylum for the Insane, Brattleboro', Vermont.

Dr. B. D. Eastman, Lunatic Hospital, Worcester, Massachusetts.

Dr. O. Everts, Hospital for the Insane, Indianapolis, Indiana.

Dr. C. C. Forbes, Central Lunatic Asylum, Anchorage, Kentucky.

Dr. W. W. Godding, Government Hospital for the Insane, Washington, District of Columbia.

Dr. John P. Gray, State Lunatic Asylum, Utica, New York.

Dr. Eugene Grissom, Insane Asylum of North Carolina, Raleigh, North Carolina.

Dr. Richard Gundry, Superintendent elect Maryland Hospital, Catonsville, Maryland.

Dr. C. H. Hughes, St. Louis, Missouri.

Dr. Walter Kempster, Northern Hospital for the Insane, Winnebago, Wisconsin.

Dr. Edwin A. Kilbourne, Northern Hospital for the Insane, Elgin, Illinois.

Dr. Thomas S. Kirkbride, Pennsylvania Hospital for the Insane, Philadelphia, Pennsylvania.

Dr. John Kirby, Assistant Physician, State Lunatic Asylum, Trenton, New Jersey.

Dr. Walter R. Langdon, Assistant Physician, Asylum for the Insane, Stockton, California.

Dr. A. E. Macdonald, City Asylum for the Insane, Ward's Island, New York.

Dr. C. S. MacDonald, State Asylum for Insane Criminals, Auburn, New York.

Dr. D. A. Morse, Asylum for the Insane, Dayton, Ohio.

Dr. Charles H. Nichols, Bloomingdale Asylum, New York.

Dr. Isaac Ray, Philadelphia, Pennsylvania.

Dr. Albert Reynolds, Hospital for the Insane, Independence, Iowa.

Dr. John W. Sawyer, Butler Hospital, Providence, Rhode Island.

Dr. S. S. Schultz, State Hospital for the Insane, Danville, Pennsylvania.

Dr. A. M. Shew, Connecticut Hospital for the Insane Middletown, Connecticut.

Dr. T. R. H. Smith, Lunatic Asylum No. 1, Fulton, Missouri.

Dr. Charles W. Stevens, St. Louis, Missouri.

Dr. William H. Stokes, Mount Hope Retreat, Baltimore, Maryland.

Dr. William W. Strew, City Lunatic Asylum, Blackwell's Island, New York.

Dr. J. Strong, Asylum for the Insane, Cleveland, Ohio.

Dr. J. D. Thomson, Mt. Hope Retreat, Baltimore, Maryland.

Dr. George C. Palmer, Asylum for the Insane, Kalamazoo, Michigan.

Dr. Clement A. Walker, Lunatic Hospital, Boston, Massachusetts.

Dr. D. R. Wallace, Hospital for the Insane, Austin, Texas.

The following gentlemen were present by invitation :

A. E. Elmore, Esq., President of the Board of Charities of Wisconsin.

D. A. Ogden, Esq., Trustee of the Willard Asylum, Willard, New York.

John T. Richardson, M. D., Commissioner of the State Hospital for the Insane, Indianapolis, Indiana.

A. P. Langworthy, M. D., of the Board of Administrators of the Asylum for the Insane of Louisiana.

———o———

The thirty-third annual meeting was held in Providence,

Rhode Island, commencing on June 10, 1879. The following members were present at the meeting :

Dr. J. B. Andrews, Assistant Physician, of the State Lunatic Asylum, Utica, New York.

Dr. J. P. Bancroft, Asylum for the Insane, Concord, New Hampshire.

Dr. C. K. Bartlett, Hospital for the Insane, St. Peter, Minnesota.

Dr. H. Black, Eastern Lunatic Asylum, Williamsburg, Virginia.

Dr. J. P. Brown, Lunatic Hospital, Taunton, Massachusetts.

Dr. P. Bryce, Hospital for the Insane, Tuscaloosa, Alabama.

Dr. D. R. Burrell, Brigham Hall, Canandaigua, New York.

Dr. John S. Butler, Hartford, Connecticut.

Dr. John H. Callender, Hospital for the Insane, Nashville, Tennessee.

Dr. J. B. Camden, Hospital for the Insane, Weston, West Virginia.

Dr. Walter Channing, Private Hospital for Insane, Brookline, Massachusetts.

Dr. John B. Chapin, Willard Asylum for the Insane, Willard, New York.

Dr. R. C. Chenault, Eastern Lunatic Asylum, Lexington, Kentucky.

Dr. Daniel Clark, Asylum for the Insane, Toronto, Ontario.

Dr. John Curwen, Pennsylvania State Lunatic Hospital, Harrisburg, Pennsylvania.

Dr. Joseph Draper, Asylum for the Insane, Brattleboro, Vermont.

Dr. J. W. Fisher, Assistant Physician of the Hospital for the Insane, Mendota, Wisconsin.

Dr. F. T. Fuller, Assistant Physician, Insane Asylum, Raleigh, North Carolina.

Dr. W. W. Godding, Government Hospital for the Insane, Washington, District of Columbia.

Dr. John C. Hall, Friends' Asylum, Frankford, Philadelphia, Pennsylvania.

Dr. W. B. Hallock, Cromwell Hall, Cromwell, Connecticut.

Dr. Henry M. Harlow, Hospital for the Insane, Augusta, Maine.

Dr. J. W. Hatch, Jr., Assistant Physician, State Asylum for the Insane, Napa, California.

Dr. Henry M. Hurd, Eastern Michigan Asylum, Pontiac, Michigan.

Dr. Walter Kempster, Northern Hospital for the Insane, Winnebago, Wisconsin.

Dr. Thomas S. Kirkbride, Pennsylvania Hospital for the Insane, Philadelphia, Pennsylvania.

Dr. W. H. Lathrop, Asylum for Chronic Insane, Tewksbury, Massachusetts.

Dr. C. S. May, State Lunatic Asylum, Danvers, Massachusetts.

Dr. Edward Mead, Boston, Massachusetts.

Dr. T. J. Mitchell, State Insane Asylum, Jackson, Mississippi.

Dr. D. A. Morse, Asylum for the Insane, Dayton, Ohio.

Dr. Charles H. Nichols, Bloomingdale Asylum, New York.

Dr. George C. Palmer, Asylum for the Insane, Kalamazoo, Michigan.

Dr. John G. Park, Worcester Lunatic Hospital, Worcester, Massachusetts.

Dr. Hosea M. Quiby, Asylum for Chronic Insane, Worcester, Massachusetts.

Dr. Joseph A. Reed, Western Pennsylvania Hospital for the Insane, Dixmont, Pennsylvania.

Dr. A. R. Reid, Nova Scotia Hospital for the Insane, Halifax, Nova Scotia.

Dr. Ira Russell, Private Asylum, Winchendon Highlands, Massachusetts.

Dr. John W. Sawyer, Butler Hospital, Providence, Rhode Island.

Dr. S. S. Schultz, State Hospital for the Insane, Danville, Pennsylvania.

Dr. A. M. Shew, Hospital for the Insane, Middletown, Connecticut.

Dr. H. P. Stearns, Retreat for the Insane, Hartford, Connecticut.

Dr. James J. Steeves, Provincial Lunatic Asylum, St. John, New Brunswick.

Dr. W. W. Strew, City Lunatic Asylum, Blackwell's Island, New York.

Dr. J. Strong, Asylum for the Insane, Cleveland, Ohio.

Dr. Clement A. Walker, Lunatic Hospital, Boston, Massachusetts.

Dr. J. M. Wallace, Asylum for the Insane, Hamilton, Ontario.

The following gentlemen were present by invitation :

Mr. Charles J. Coffin, of the Board of Charities of Massachusetts.

Mr: George W. Jones, Trustee of the Willard Asylum.

Dr. George Brown, Barre, Massachusetts.

Dr. Allen M. Lane, Hamilton, New York.

Dr. Caswell, President Rhode Island Medical Society.

Dr. Anthony, President of Providence Medical Association.

Dr. Gurden W. Russell, Hartford, Connecticut.

Dr. J. F. Noyes, Detroit, Michigan.

Dr. Theo. W. Fisher, Boston, Massachusetts.

————o————

The thirty-fourth annual meeting was held in Philadelphia, commencing on Tuesday, May 25, 1880. The following members were present :

Dr. J. K. Bauduy, St. Vincent's Institution for the Insane, St. Louis, Missouri.

Dr. D. T. Boughton, State Hospital for the Insane, Mendota, Wisconsin.

Dr. J. P. Brown, State Lunatic Hospital, Taunton, Massachuchusettss.

Dr. P. Bryce, Alabama Insane Hospital, Tuscaloosa, Alabama.

Dr. R. M. Bucke, Asylum for the Insane, London, Ontario.

Dr. D. R. Burrell, Brigham Hall, Canandaigua, New York.

Dr. H. A. Buttolph, State Asylum for the Insane, Morris Plains, New Jersey.

Dr. John H. Callender, Hospital for the Insane, Nashville, Tennessee.

Dr. T. B. Camden, Hospital for the Insane, Weston, West Virginia.

Dr. John B. Chapin, Willard Asylum, Willard, New York.

Dr. Daniel Clark, Asylum for the Insane, Toronto, Ontario.

Dr. H. F. Carriel, Hospital for the Insane, Jacksonville, Illinois.

Dr. John Curwen, Pennsylvania State Lunatic Asylum, Harrisburg, Pennsylvania.

Dr. Theo. Dimon, Asylum for Insane Criminals, Auburn, New York.

Dr. B. D. Eastman, Insane Asylum, Topeka, Kansas.

Dr. Orpheus Everts, Cincinnati Sanitarium, College Hill, Ohio.

Dr. F. T. Fuller, Assistant Physician of Insane Asylum, Raleigh, North Carolina.

Dr. W. W. Godding, Government Hospital for the Insane, Washington, District of Columbia.

Dr. John P. Gray, State Lunatic Asylum, Utica, New York.

Dr. Richard Gundry, Maryland Hospital for the Insane, Catonsville, Maryland.

Dr. John C. Hall, Friends' Asylum, Frankford, Philadelphia, Pennsylvania.

Dr. Henry M. Hurd, Eastern Michigan Asylum, Pontiac, Michigan.

Dr. Walter Kempster, Northern Hospital for the Insane, Winnebago, Wisconsin.

Dr. Thomas S. Kirkbride, Pennsylvania Hospital for the Insane, Philadelphia, Pennsylvania.

Dr. A. E. Macdonald, City Lunatic Asylum, Ward's Island. New York.

Dr C. F. Macdonald, Binghamton Asylum for the Insane, Binghamton, New York.

Dr. S. B, McGlumphy, Hospital for the Insane, Yankton, Dakota.

Dr. C. S. May, Lunatic Hospital, Danvers, Massachusetts.

Dr. W. G. Metcalf, Asylum for the Insane, Kingston, Ontario.

Dr. C. A. Miller, Longview Asylum, Carthage, Ohio.

Dr. D. A. Morse, Asylum for the Insane, Dayton, Ohio.

Dr. Charles H. Nichols, Bloomingdale Asylum, New York.

Dr. George C. Palmer, Asylum for the Insane, Kalamazoo, Michigan.

Dr. J. O. Powell, Insane Asylum, Milledgeville, Georgia.

Dr. Isaac Ray, Philadelphia, Pennsylvania.

Dr. Joseph A. Reed, Western Pennsylvania Hospital for the Insane, Dixmont, Pennsylvania.

Dr. D. D. Richardson, State Hospital for the Insane, Warren, Pennsylvania.

Dr. Joseph G. Rogers, Hospital for the Insane, Indianapolis, Indiana.

Dr. John W. Sawyer, Butler Hospital, Providence, Rhode Island.

Dr. S. S. Schultz, State Hospital for the Insane, Danville, Pennsylvania.

Dr. G. A. Shurtleff, Asylum for the Insane, Stockton, California.

Dr. James T. Steeves, Provincial Lunatic Asylum, St. John, New Brunswick.

Dr. J. Strong, Asylum for the Insane, Cleveland, Ohio.

Dr. J. D. Thompson, Mt. Hope Retreat, Baltimore, Maryland.

Dr. Clement A. Walker, Lunatic Hospital, Boston, Massachusetts.

Dr. John W. Ward, State Lunatic Asylum, Trenton, New Jersey.

Dr. H. Wardner, Hospital for the Insane, Anna, Illinois.

Dr. J. H. Worthington, Baltimore, Maryland.

Dr. John S. Woodside, Assistant Physician, Kings County Lunatic Asylum, Flatbush, New York.

The following gentlemen were present by invitation :

Dr. Alfred T. Livingston, Philadelphia, Pennsylvania.

Dr. J. N. Kerlin, Institution for Feeble Minded Children, Media, Pennsylvania.

Mr. Gardner A. Churchill, Trustee of the Lunatic Hospital, Danvers, Massachusetts.

Mr. George W. Jones, Trustee of the Willard Asylum, Willard, New York.

Dr. Traill Green, Trustee of the Pennsylvania State Lunatic Hospital, Harrisburg, Pennsylvania.

Dr. William Corson, Commissioner of the State Hospital for the Insane, Warren, Pa.

John C. Allen and Henry Haines, Managers of the Friends' Asylum, Frankford, Philadelphia, Pennsylvania.

Dr. D. Hack Tuke was elected an Honorary member of the Association.

———o———

The thirty-fifth annual meeting was held in Toronto, Ontario, commencing on Tuesday, June 14, 1881. The following members were present :

Dr. J. B. Andrews, State Asylum for the Insane, Buffalo, New York.

Dr. R. Barksdale, Central Lunatic Asylum, Richmond, Virginia.

Dr. J. W. Barstow, Sanford Hall, Flushing, New York.

Dr. H. Black, Eastern Lunatic Asylum, Williamsburg, Virginia.

Dr. R. M. Bucke, Asylum for the Insane, London, Ontario.

Dr. W. O. Bullock, Eastern Lunatic Asylum, Lexington, Kentucky.

Dr. D. R. Burrell, Brigham Hall, Canandaigua, New York.

Dr. A. P. Busey, Assistant Physician, Lunatic Asylum, No. 2, St. Joseph, Missouri.

Dr. John H. Callender, Hospital for the Insane, Nashville, Tennessee.

Dr. H. F. Carriel, Central Hospital for the Insane, Jacksonville, Illinois.

Dr. Daniel Clark, Asylum for the Insane, Toronto, Ontario.

Dr. John Curwen, Harrisburg, Pennsylvania.

Dr. James H. Denny, Boston, Massachusetts.

Dr. R. S. Dewey, Eastern Hospital for the Insane, Kankakee, Illinois.

Dr. Orpheus Everts, Cincinnati Sanitarium, College Hill, Ohio.

Dr. A. M. Fauntleroy, Western Lunatic Asylum, Staunton, Virginia.

Dr. Theodore W. Fisher, Lunatic Hospital, Boston, Massachusetts.

Dr. T. M. Franklin, City Lunatic Asylum, Blackwell's Island, New York.

Dr. R. H. Gale, Central Kentucky Lunatic Asylum, Anchorage, Kentucky.

Dr. J. Z. Gerhard, Pennsylvania State Lunatic Hospital, Harrisburg, Pennsylvania.

Dr. William B. Goldsmith, Lunatic Hospital, Danvers, Massachusetts.

Dr. John P. Gray, State Lunatic Asylum, Utica, New York.

Dr. Richard Gundry, Maryland Hospital for the Insane, Catonsville, Maryland.

Dr. C. H. Hughes, St. Louis, Missouri.

Dr. Henry M. Hurd, Eastern Michigan Asylum, Pontiac, Michigan.

Dr. A. E. Macdonald, City Lunatic Asylum, Ward's Island, New York.

Dr. H. P. Mathewson, Hospital for the Insane, Lincoln, Nebraska.

Dr. W. G. Metcalf, Asylum for the Insane, Kingston, Ontario.

Dr. Charles A. Miller, Longview Asylum, Carthage, Ohio.

Dr. J. A. Reed, Western Pennsylvania Hospital for the Insane, Dixmont, Pennsylvania.

Dr. A. P. Reid, Hospital for the Insane, Halifax, Nova Scotia.

Dr. Joseph G. Rogers, Hospital for the Insane, Indianapolis, Indiana.

Dr. F. E. Roy, Lunatic Asylum, Quebec.

Dr. H. C. Rutter, Asylum for the Insane, Columbus, Ohio.

Dr. John W. Sawyer, Butler Hospital, Providence, Rhode Island.

Dr. J. Strong, Asylum for the Insane, Cleveland, Ohio.

Dr. B. R. Thombs, State Asylum for the Insane, Pueblo, Colorado.

Dr. H. A. Tobey, Asylum for the Insane, Dayton, Ohio.

Dr. J. M. Wallace, Asylum for the Insane, Hamilton, Ontario.

Dr. Joseph Workman, Toronto, Ontario.

The following gentlemen were present by invitation :

Mr. J. W. Langmuir, Inspector of Asylums and Prisons of Ontario.

Mr. D. A. Ogden, Trustee of the Willard Asylum, Willard, New York.

Mr. W. P. Townsend, Manager of the Western Pennsylvania Hospital for the Insane, Dixmont, Pennsylvania.

Dr. Godfrey, Trustee of the Asylum for the Insane, Dayton, Ohio.

Dr. Fulton, Editor of the Canada *Lancet*, and Professor in Trinity Medical College.

Dr. William Canniff, President of the Medical Association of the Dominion of Canada.

Dr. Graham, of the Senate of the University of Ontario.

Dr. Grant, of Ottawa, Member of the Medical Council.

Dr. A. H. Beaton, of the Orilia Asylum for Idiots.

The following gentlemen were elected Honorary members of the Association :

Dr. C. Lockhart Robertson, of England.

Dr. A. Motet, of Paris.

Dr. A. Tamburini, of Italy.

Dr. T. S. Clouston, of Scotland.

————o————

The thirty-sixth annual meeting of the Association was held in Cincinnati, Ohio, commencing on Tuesday, May 30, 1882. The following members were present :

Dr. J. B. Andrews, State Asylum for the Insane, Buffalo, New York.

Dr. W. J. Bland, Hospital for the Insane, Weston, West Virginia.

Dr. R. M. Bucke, Asylum for the Insane, London, Ontario.

Dr. D. R. Burrell, Brigham Hall, Canandaigua, New York.

Dr. John H. Callender, Hospital for the Insane, Nashville, Tennessee.

Dr. John B. Chapin, Willard Asylum, Willard, New York.

Dr. Edward Cowles, McLean Asylum, Somerville, Massachusetts.

Dr. John Curwen, State Hospital for the Insane, Warren, Pennsylvania.

Dr. R. S. Dewey, Eastern Hospital for the Insane, Kankakee, Illinois.

Dr. Orpheus Everts, Cincinnati Sanitarium, College Hill, Ohio.

Dr. Theodore W. Fisher. Lunatic Hospital, Boston, Massachusetts.

Dr. R. H. Gale, Central Lunatic Asylum, Anchorage, Kentucky.

Dr. William B. Goldsmith, Lunatic Hospital, Danvers, Massachusetts.

Dr. L. J. Graham, State Lunatic Asylum, Austin, Texas.

Dr. John P. Gray, State Lunatic Asylum, Utica, New York.

Dr. Eugene Grissom, Insane Asylum, Raleigh, North Carolina.

Dr. Richard Gundry, Maryland Hospital, Catonsville, Maryland.

Dr. John C. Hall, Friends' Asylum, Frankford, Philadelphia, Pennsylvania.

Dr. F. W. Hatch, Jr., Assistant Physician, Asylum for the Insane, Napa, California.

Dr. Charles H. Hughes, St. Louis, Missouri.

Dr. Henry M. Hurd, Eastern Michigan Asylum, Pontiac, Michigan.

Dr. E. A. Kilbourne, Northern Hospital for the Insane, Elgin, Illinois.

Dr. Andrew McFarland, Oak Lawn Retreat, Jacksonville, Illinois.

Dr. H. P. Mathewson, State Hospital for the Insane, Lincoln, Nebraska.

Dr. C. A. Miller, Longview Asylum, Carthage, Ohio.

Dr. T. J. Mitchell, Lunatic Asylum, Jackson, Mississippi.

Dr. A. R. Moulton, Assistant Physician, Lunatic Hospital, Worcester, Massachusetts.

Dr. Charles H. Nichols, Bloomingdale Asylum, New York City.

Dr. George C. Palmer, Asylum for the Insane, Kalamazoo, Michigan.

Dr. Joseph A. Reed, Western Pennsylvania Hospital for the Insane, Dixmont, Pennsylvania.

Dr. A. B. Richardson, Asylum for the Insane, Athens, Ohio.

Dr. Joseph G. Rogers, Hospital for the Insane, Indianapolis, Indiana.

Dr. John W. Sawyer, Butler Hospital, Providence, Rhode Island.

Dr. S. S. Schultz, State Hospital for the Insane, Danville, Pennsylvania.

Dr. Henry P. Stearns, Retreat for the Insane, Hartford, Connecticut.

Dr. Charles W. Stevens, St. Louis, Missouri.

Dr. J. Strong, Asylum for the Insane, Cleveland, Ohio.

Dr. H. A. Toby, Asylum for the Insane, Dayton, Ohio.

Dr. J. M. Wallace, Asylum for the Insane, Hamilton, Ontario.

Dr. H. Wardner, Hospital for the Insane, Anna, Illinois.

Dr. James M. Whitaker, Assistant Physician, Lunatic Asylum, Milledgeville, Georgia.

The following resolution was adopted :

Resolved, That the usage of the Association, in respect to the tenure of the office of President and Vice President of this body, be so far changed that hereafter there shall be elected a President and Vice President, to hold their respective offices for a period of one year, and that the President present an annual address, which shall be deemed exempt from critical discussion, unless the Association shall direct otherwise.

Dr. C. A. Walker, having resigned the office of President, Dr. John H. Callender was elected President, and Dr. John P. Gray, Vice President, under the resolution adopted at this meeting.

The following resolution was also adopted :

Resolved, That on the last day of each annual meeting of the Association of Medical Superintendents of American Institutions for the Insane, the President shall appoint committees, whose duty it shall be to report at the next annual meeting upon the state and progress of the various important divisions of special science and art, relating to the insane as assigned to and accepted by them, and whose chairmen shall be ethically responsible for the proper presentation of such reports.

These subjects were subsequently divided as follows :

1. Annual Necrology of the Association.
2. Cerebro-Spinal Physiology.
3. Cerebro-Spinal Pathology.
4. Therapeutics of Insanity.
5. Bibliography of Insanity.
6. Relation of Eccentric Diseases to Insanity.
7. Asylum Location, Construction and Sanitation.
8. On Medico-Legal relations of the Insane.
9. On the Treatment of Insanity.

The thirty-seventh annual meeting of the Association was held at Newport, Rhode Island, commencing on June 26, 1883. The following members were present :

Dr. J. B. Andrews, State Asylum for the Insane, Buffalo, New York.

Dr. J. P. Bancroft, Concord, New Hampshire.

Dr. J. W. Barstow, Sanford Hall, Flushing, New York.

Dr. W. J. Bland, State Hospital for the Insane, Weston, West Virginia.

Dr. J. P. Brown, Lunatic Hospital, Taunton, Massachusetts.

Dr. John H. Callender, State Hospital for the Insane, Nashville, Tennessee.

Dr. George C. Catlett, Lunatic Asylum No. 2, St. Joseph, Missouri.

Dr. John B. Chapin, Willard Asylum, Willard, New York.

Dr. R. H. Chase, State Hospital for the Insane, Norristown, Pennsylvania.

Dr. Daniel Clark. Asylum for the Insane, Toronto, Ontario.

Dr. Edward Cowles, McLean Asylum, Somerville, Massachusetts.

Dr. John Curwen, State Hospital for the Insane, Warren, Pennsylvania.

Dr. James H. Denny, Boston, Massachusetts.

Dr. Joseph Draper, Asylum for the Insane, Brattleboro', Vermont.

Dr. O. Everts, Cincinnati Sanitarium, College Hill, Ohio.

Dr. Theo. W. Fisher, Lunatic Hospital, Boston, Massachusetts.

Dr. J. M. Franklin, City Lunatic Asylum, Blackwell's Island, New York.

Dr. R. H. Gale, Central Lunatic Asylum, Anchorage, Kentucky.

Dr. J. Z. Gerhard, Pennsylvania State Lunatic Hospital, Harrisburg, Pennsylvania.

Dr. W. W. Godding, Government Hospital for the Insane. Washington, District of Columbia.

Dr. W. B. Goldsmith, Lunatic Hospital, Danvers, Massachusetts.

Dr. John P. Gray, State Lunatic Asylum, Utica, New York.

Dr. John C. Hall, Friend's Asylum, Frankford, Philadelphia, Pennsylvania.

Dr. W. B. Hallock, Cromwell Hall, Cromwell, Connecticut.

Dr. Charles J. Hill, Assistant Physician, Mount Hope Retreat, Baltimore, Maryland.

Dr. G. H. Hill, Hospital for the Insane, Independence, Iowa.

Dr. Henry M. Hurd, Asylum for the Insane, Pontiac, Michigan.

Dr. George F. Jelly, Boston, Massachusetts.

Dr. A. E. Macdonald, City Lunatic Asylum, Ward's Island, New York.

Dr. W. G. Metcalf, Kingston, Ontario.

Dr. Charles H. Nichols, Bloomingdale Asylum, New York.

Dr. George C. Palmer, Asylum for the Insane, Kalamazoo, Michigan.

Dr. H. M. Quinby, Asylum for Chronic Insane, Worcester, Massachusetts.

Dr. A. B. Richardson, Asylum for the Insane, Athens, Ohio.

Dr. Joseph G. Rogers, Hospital for the Insane, Indianapolis, Indiana.

Dr. F. E. Roy, Lunatic Asylum, Quebec, Canada.

Dr. Ira Russell, Winchendon, Massachusetts.

Dr. John W. Sawyer, Butler Hospital, Providence, Rhode Island.

Dr. S. S. Schultz, State Hospital for the Insane, Danville, Pennsylvania.

Dr. A. M. Shew, Hospital for the Insane, Middletown, Connecticut.

Dr. H. P. Stearns, Retreat for the Insane, Hartford, Connecticut.

Dr. J. T. Steeves, Provincial Lunatic Asylum, St. John, New Brunswick.

Dr. G. B. Twitchell, Keene, New Hampshire.

The following gentlemen were present by invitation :

Dr. Horatio R. Storer, President of the Newport Medical Society.

Dr. Foster Pratt, Manager of the Asylum for the Insane, Kalamazoo, Michigan.

Mr. D. A. Ogden, Manager of the Willard Asylum, Willard, New York.

Rev. Mr. Willard, Secretary of the Trustees of the Hospital for the Insane, Middletown, Connecticut.

Dr. A. G. Watson, Newport, Rhode Island.

Mr. George Gordon King, Newport, Rhode Island.

Mr. A. G. Barstow, President of the Board of Trustees of Butler Hospital, Providence, Rhode Island.

Mr. Brownell, Trustee of the Butler Hospital.

Dr. Theodore Meynert, of Vienna, was elected an Honorary member of the Association.

———o———

The thirty-eighth annual meeting was held at the Continental Hotel, in the city of Philadelphia, commencing at 10 A. M., of May 13, 1884. The following members were present during the sessions :

J. B. Andrews, M. D., Asylum for the Insane, Buffalo, New York.

J. P. Bancroft, M. D., Concord, New Hampshire.

W. J. Bland, M. D., Hospital for the Insane, Weston, West Virginia.

J. P. Brown, M. D., Lunatic Hospital, Taunton, Massachusetts.

W. T. Browne, M. D., Asylum for the Insane, Stockton, California.

R. M. Bucke, M. D., Asylum for the Insane, London, Ontario.

D. R. Burrell, M. D., Brigham Hall, Canandaigua, New York.

John H. Callender, M. D., Hospital for the Insane, Nashville, Tennessee.

H. F. Carriel, M. D., Hospital for the Insane, Jacksonville, Illinois.

George T. Catlett, M. D., Lunatic Asylum No. 2, St. Joseph, Missouri.

Walter Channing, M. D., Brookline, Massachusetts.

John B. Chapin, M. D., Asylum for the Insane, Willard, New York.

R. H. Chase, M. D., State Hospital for the Insane, Norristown, Pennsylvania.

Edward Cowles, M. D., McLean Asylum, Somerville, Massachusetts.

John Curwen, M. D., State Hospital for the Insane, Warren, Pennsylvania.

A. N. Denton, M. D., Asylum for the Insane, Austin, Texas.

R. S. Dewey, M. D., Eastern Hospital for the Insane, Kankakee, Illinois.

Pliny Earle, M. D., Lunatic Hospital, Northampton, Massachusetts.

Orpheus Everts, M. D., Cincinnati Sanitarium, College Hill, Ohio.

Theodore W. Fisher, M. D., Lunatic Hospital, Boston, Massachusetts.

T. M. Franklin, M. D., City Lunatic Asylum, Blackwell's Island, New York.

J. Z. Gerhard, M. D., Pennsylvania State Lunatic Hospital, Harrisburg, Pennsylvania.

W. W. Godding, M. D., Government Hospital for the Insane, Washington, District of Columbia.

John P. Gray, M. D., State Lunatic Asylum, Utica, New York.

Eugene Grissom, M. D., Insane Asylum, Raleigh, North Carolina.

John C. Hall, M. D., Friends' Asylum for the Insane, Frankford, Philadelphia, Pennsylvania.

Charles J. Hill, M. D., Assistant Physician, Mt. Hope Retreat, Baltimore, Maryland.

S. Preston Jones. M. D., Department for Males, Pennsylvania Hospital for the Insane, Philadelphia, Pennsylvania.

Alfred T. Livingston, M. D., Wa-Wa, Delaware County, Pennsylvania.

P. L. Murphy, M. D., Western North Carolina Insane Asylum, Morganton, North Carolina.

Charles H. Nichols, M. D., Bloomingdale Asylum, New York.

George C. Palmer, M. D., Asylum for the Insane, Kalamazoo, Michigan.

J. Willoughby Phillips, M. D., Assistant Physician, Burn Brae, Delaware County, Pennsylvania.

T. O. Powell, M. D., Asylum for the Insane, Milledgeville, Georgia.

A. B. Richardson, M. D., Asylum for the Insane, Athens, Ohio.

D. D. Richardson, M. D., Department for the Insane, Almshouse, Philadelphia, Pennsylvania.

J. D. Roberts, M. D., Eastern North Carolina Insane Asylum, Goldsboro, North Carolina.

Ira Russell, M. D., Highlands, Winchendon, Massachusetts.

John W. Sawyer, M. D., Butler Hospital, Providence, Rhode Island.

S. S. Schultz, M. D., State Hospital for the Insane, Danville, Pennsylvania.

A. M. Shew, M. D., Hospital for the Insane, Middletown, Connecticut.

George S. Sinclair, M. D., Assistant Physician, Hospital for the Insane, Halifax, Nova Scotia.

E. E. Smith, M. D., Assistant Physician, Asylum for the Insane, Morris Plains, New Jersey.

Henry P. Stearns, M. D., Retreat for the Insane, Hartford, Connecticut.

James T. Steeves, M. D., Provincial Lunatic Asylum, St. John, New Brunswick.

J. Strong, M. D., Asylum for the Insane, Cleveland, Ohio.

H. A. Toby, M. D., Asylum for the Insane, Dayton, Ohio.

George B. Twitchell, M. D., Keene, New Hampshire.

142

J. M. Wallace, M. D., Asylum for the Insane, Hamilton, Ontario.

John W. Ward, M. D., State Lunatic Asylum, Trenton, New Jersey.

Dr. Curwen offered the following resolution which was unanimously adopted on a rising vote :

Resolved, That in the death of our fellow member, Dr. Thomas S. Kirkbride, this Association has lost one of its ablest members, who, during the whole period of its existence, had given to it most earnest and devoted thought and attention, and whose counsels were always wise, cautious and most enlightened.

A kind, warm-hearted and sympathizing friend, a faithful and prudent counsellor, a genial and cheerful companion, and a most able, laborious and devoted Physician and Superintendent, no one who was privileged to know him in those relations can fail to feel the great blank which has been made by his removal.

Privileged to continue in active continuous service longer than any other member, his latest thoughts were given to the consideration of those things which would most benefit those for whom for more than forty years he had thought and labored.

Dr. John P. Gray delivered the address as President and then introduced his successor, Dr. Pliny Earle.

This being the fortieth year from the formation of the Association, in accordance with a previous arrangement made last year, addresses were delivered on the History of the Association and its necrology by Dr. John Curwen, on progress in the treatment of the Insane by Dr. Henry P. Stearns, on progress in provision for the Insane by Dr. W. W. Godding.

The Association in a body visited the Medical Society of the State of Pennsylvania in session, and also attended the reception of the President of that Society in the evening.

Dr. Foster Pratt, one of the Trustees of the Asylum for the Insane at Kalamazoo, Michigan, offered the following resolutions which were adopted :

WHEREAS, By a comparison of the statistics of the "defective" classes of our population," as shown by the eighth, ninth and tenth Census, it appears

First, That the proportion of insane to total population in the United States is rapidly increasing, and Secondly, That a prominent factor in this increase is the large defective element found among the "Foreign born" who have emigrated to us since 1847 and 1848—an element which now constitutes one-eighth of our total population, but which furnishes one-third of its paupers, one-third of its criminals, and one-third of its insane, and WHEREAS, While the cost of buildings to suitably keep and the annual tax to properly maintain these classes fall wholly and heavily on the several States and Territories, they are inhibited by a national law from enacting and enforcing effective measures to prevent or to mitigate these evils, so far as they are caused by immigration, now therefore

Resolved, That the Association of Medical Superintendents of American Institutions for the Insane respectfully urges the Congress of the United States to give early and earnest attention to this important subject, to the end that emigration laws may be enacted by it which, while they do not unreasonably obstruct the immigration of healthy and self-dependent persons, will effectively prevent the emigration and the exportation to our ports of the so-called defective classes of Europe and Asia.

Resolved. That in furtherance of this object a copy of these Resolutions and Preamble be forwarded by the President and Secretary of this Association to the President of the United States, and to the President of the Senate and Speaker of the House of Representatives at Washington, for consideration by them and by Congress; also to the Governor and the Presiding officers of the Legislature of each State of the Union, that they and the people they severally represent, who are most affected by the pecuniary burdens and by the vital and moral evils caused by an unrestricted and unregulated immigration, may be moved to take such action as they deem best to secure early and efficient action by Congress (with whom alone is the power) to abate the great and growing evils to which public attention is hereby called.

Resolved, That a copy of these Resolutions and Preamble be also sent to the Secretary of each medical society in the several States, with the request that the medical profession generally unite with us in the attempt to obtain the required remedy for these great evils.

BIOGRAPHICAL NOTICES.

————o————

John E. Tyler, M. D., by Isaac Ray, M. D. :

My relations with Dr. Tyler were not of the kind that bring
to view all the sides and aspects of a man's nature, but they suf-
ficed to reveal to me many sterling qualities well worthy of the
highest esteem. Starting with his mind well prepared by a college training, and a faithful study of his profession, he obtained
in due season the merited reward of such preparation. While
engaged in a general practice, embracing to a large extent the
most respectable and cultivated part of the community, he was
selected by the Trustees of the State Asylum of New Hampshire
to become its Superintendent. So well did he discharge this
trust, that, under his charge, the institution notably prospered,
while he established his own reputation in this peculiar calling.
On the death of Dr. Booth, the Superintendent of the McLean
Asylum, the Trustees of that institution had little hesitation in
making Dr. Tyler his successor. Here the best qualities of his nature were brought into action, as they never had been before, and
his remarkable fitness for the kind of duty he had assumed was
admirably displayed. In no other similar institution in the country are larger drafts made on the patience, the temper, the indus-
try, the zeal, in short, on all the moral and intellectual resources
of the Superintendent. For thirteen years he stood the trial,
steadily gaining the approbation of his Trustees, the confidence
and esteem of his patients, and the respect of his medical brethren. He came to the work with a correct appreciation of its responsibilities, and an earnest endeavor to achieve the highest
measure of success. Thenceforth it became the all-absorbing
interest of his life. Surrounded by memorials of his predeces-
sors, he needed no other incentive to make himself worthy a

place by the side of a Booth, a Bell, a Lee and a Wyman. It was a purpose worthy of the noblest ambition. How worthily he achieved it, we learn from the abundant testimony both of his patients and his employers. He cared little for popular applause, and was well satisfied with the approbation of those who, alone, were the proper judges of his merits.

He had many qualities indispensable to success in his calling. Without any profound study of psychological science, he possessed that nice discernment of abnormal mental conditions which springs from a happy faculty of observation,—a faculty which may be improved by use, but is chiefly a gift of nature. It enabled him to look beneath the surface, and discern signs of irregular action that would escape the notice of others less happily endowed. His success was much promoted by a genial temper and a pleasing address that always made him a welcome companion, bringing, at every visit, a gleam of sunshine to many a darkened soul. Few could resist the cheering influence of his hearty laugh and pleasant words, well-timed and skilfully expressed as they always were.

In the character of expert in cases of insanity, in which he often appeared, it would be no small praise to say that he did no discredit to his profession, but he also did something more. He was always cool, self-collected, not easily embarrassed, and was unusually successful in obtaining respect and confidence for his statements. He soon learned, what some experts never learn at all, that to satisfy himself of the correctness of his positions is scarcely more important than to forsee how they will strike others. It is this kind of prescience which makes one sure that the ground he takes is tenable, and enables him to anticipate the assaults he will have to meet. After a service of thirteen years his health had received such a shock from a malarial fever contracted while on a visit South, that he felt obliged to resign, and seek the restorative influences of a prolonged stay in Europe. On his return, with his condition greatly improved, he engaged in private practice and soon had all the employment he desired. He had been appointed while in the Asylum, Professor of Nervous Diseases in Harvard University, and the last professional act he did was to give the usual lecture of his course. He will be much missed in that community, for he was widely known and esteemed, and in various relations his counsel was sought for and highly prized.

As a member of this Association his presence among us always met with a hearty welcome. His words were ever wise and timely. He was not much inclined to writing or speaking, but when he did write or speak, it was something well worth listening to.

———

Dr. George Syng Bryant was born in Old Virginia, in 1825, and died in June, 1875, in full vigor of manhood. He was educated at Hampden Sydney College, and graduated at an early age, it is said, with the honors of his class ; studied medicine and graduated from Old Jefferson, in Philadelphia, in 1845. Soon after he removed to Mississippi, where he practiced his profession very successfully for about ten years, up to the commencement of the late civil war, when he was appointed a surgeon in the Confederate service, and won for himself distinction in that service. At the close of the war he removed to St. Louis, Missouri, but was induced to leave that place on account of failing health, brought about by exposure during the war. He removed to Lexington, Kentucky, in the spring of 1868 ; soon made for himself a reputation as a man of more than ordinary ability, became an active and prominent member of the Kentucky State Medical Society, and won the exalted esteem of the profession generally throughout the State. His enthusiasm for his profession, his admiration for the masters of his science, his studious habits, and his contributions to the various medical journals, all marked him as a man of no ordinary cast. As a gynæcologist he was distinguished in the west, especially as an operator and also as an inventor.

With those with whom he was associated in the management of the Eastern Kentucky Asylum, from the highest to the lowest, all continue to speak of his uniform kindness and his unceasing efforts to make every one around him comfortable and happy.

He will be missed, indeed, from our Association, from the Kentucky State Medical Society, to which he was a contributor, from the profession generally where he lived, and among whom he had many warm admirers, and from society generally.

———

John Waddell, whose father was a native of Shotts, Scotland, was born in Truro, Nova Scotia, on March 17, 1810. He was the youngest son of the Reverend John Waddel, an eminent Presbyterian clergyman, and brother of the late James Waddell, also a distinguished member of the Presbyterian church. The early part of his education was received at the Grammar School in Truro : subsequently he attended the Pictou Academy, where he spent several years, completing a full course of liberal culture. At the end of this period he engaged in business, continuing for one year, but finding this enterprise uncongenial it was abandoned. In the year 1834, he commenced the study of medicine in his native place, under the preceptorship of Dr. Lynd. He next proceeded to Glasgow, continuing his medical studies there, and on the 18th of October, 1839, he received his diploma from the Royal College of Surgeons, London. After obtaining his degree, the doctor attended medical lectures in Paris during the winter of 1839 and 184c. In the summer of 1840, he returned to Truro, Nova Scotia, and entered upon the practice of his chosen profession. During the following nine years he was engaged in general practice, and, being eminently successful, he extended his name and fame far beyond the immediate sphere of his labors. In 1849, Dr. Waddell was appointed Medical Superintendent of the Provincial Lunatic Asylum at St. John, New Brunswick, and in December of that year, he entered upon the duties connected therewith. In the management of this institution, the doctor found a sphere congenial to his order of mind, and he soon won a reputation more than provincial. In a preeminent degree he possessed the qualities of mind and heart to insure success in his chosen field. His administrative ability was of a high order, he was prudent, practical and economical in his management, and, averse to the use of too definitely written rules, preferring a frequent resort to himself as the source of authority in the house which he controlled. His fine personnel, gentlemanly bearing, suave manners and cheerful disposition, gained for him at once the confidence and esteem of his associates, and the public as well. Whilst Dr. Waddell was urbane, generous and forgiving, yet he possessed great firmness of character ; when opposed in his cherished views or plans, his opponent found "a foeman worthy of his steel." Dr. Waddell continued Superintendent of the asylum at St. John, from December, 1849, until the 1st of May, 1875, a period extending upwards of twenty-six years, and during all that time he labored with great assiduity, and with marked success in the medical treatment of the patients, the general management of the house, and in all that pertained to the prosperity of the institution.

Far the best part of his life was devoted to a noble purpose, caring for the helpless and insane, going in and out among them at all hours of the day and night, ministering to their diseased bodies and minds, performing the office of a faithful physician. Early in the history of this Association, Dr. Waddell became an active member, taking a deep interest in its work and earnestly promoting its welfare. His agreeable social qualities, varied information and practical good sense, made him a great favorite among the members of the Association. On the doctor's retiring from the Superintendency of the asylum, he again took up his residence at Truro, his birthplace, where he himself, and his friends, hoped he might enjoy many years of quiet and peace, after his arduous life duties had been so well performed.

But this hope was not realized ; the good doctor had almost finished his course ; he had well nigh fallen before his armor was removed. The watching, the anxiety too long continued without sufficient aid, had so wrought upon his physical system and mind, that a nervous affection fastened upon him to which he soon succumbed.

On Thursday, the 29th of August, 1878, our friend, a Christian gentleman, passed away peacefully to his rest and his reward.

Dr. Thomas F. Green was born in Beaufort, S. C., on the 25th of December, 1804. He died in Midway, Georgia, on the 13th of February, 1879, of apoplexy, while Superintendent of the Georgia Lunatic Asylum. His parents were of the best class of Irish people. His father, a warm-hearted, highly educated, enthusiastic young Irish patriot, joined in the ill-fated rebellion of 1798, was forced to flee the country ; his wife, who was a Fitzgerald, a lady of noble blood, came with him to America. He had no fortune save his talents, no friends save those he won by his virtues. He began to teach, and as a teacher came to Beaufort, S. C. Here his eldest son, Thomas Fitzgerald, was born. He removed to Savannah, Georgia, where he taught a high school and was then elected a Professor in the Georgia University at Athens. He afterwards removed to Milledgeville, the capital of Georgia, and here the son was educated. He was past his majority when he studied medicine and began to practice. He located in Milledgeville and was doing well as a physician when the current of his life was changed, and turned into a direction which was to be full of blessings to his race. A northern philanthropist, who was interested in the welfare of the insane, visited Milledgeville to suggest and advocate the establishment of an asylum for them. He called a meeting of a few gentlemen of broad views and generous hearts and laid his plans before them. The warm heart of Dr. Thomas F. Green became much interested in the great question presented, and he gave it close attention. He was connected with the first effort made to secure the grant from the Legislature. In 1846 he succeeded Dr. Cooper as Superintendent of the Asylum. He continued in the office for thirty-three years. It was very small when he took hold of it. It became a grand institution, one of the largest in the Southern States, when he was called by death from it.

Dr. Green in person was short, stout, of broad, grand, humane countenance, in his youth handsome, and in his old age venerable. He was full of life, cheerful, merry, courteous, considerate. He was a sincere Christian, in his home life a model, one of the most benevolent and unselfish of men. He was devoted to the Institution ; he literally lived for the Asylum. He thought

of it, talked of it all the time. His success in the management
of it was marvellous and the blessed results of his work cannot
be told in time. He was a delightful companion, a true and
sympathizing friend, a man whom all loved, and one worthy of
all the honor heaped upon him. The moral grandeur of his
character was best illustrated by the interest he manifested in the
unfortunate.

Wm. S. Chipley, son of Rev. Stephen and Amelia Stout Chipley, was born at Lexington, Kentucky, Oct. 18, 1810, being the
third child of his parents. He inherited from both parental
lives a firm and vigorous organization, including a large and
well-balanced brain with great intellectual potentialities. He
received an academical education, and afterward the degree of
Doctor of Medicine (1832) from the Transylvania University;
which for many years was the most prominent educational establishment west of the Allegheny mountains. Soon after graduating he commenced the practice of medicine in Columbus,
Georgia, where, with the ardor of youthful ambition, and a consciousness of more than ordinary intellectual endowment, he
developed not only a great interest in, but a capacity for politics
and general affairs; but returned to Lexington, Kentucky in
1844, and limited his activities to the cultivation and practice of
his profession. Successful as a practitioner and reputable as a man
of learning. he was elected to the chair of Theory and Practice
of Medicine in the medical department of Transylvania in 1853,
and delivered the lectures from that chair until called to the position of Superintendent of the Eastern Lunatic Asylum of
Kentucky, at Lexington, in 1855, then the oldest and, perhaps, largest public provision for the insane in the west;
which position he occupied continuously for fifteen years, dis-
charging the incumbent duties with ability and marked devotion to the interests of the insane as well as the interests of
the State. The asylum was greatly enlarged by new struc-
tures, under his supervision, and ranked deservedly well among

the institutions of its class in America. Personal and political exigencies compelled Dr. Chipley to resign his superintendency in 1870—soon after which he opened a private hospital for the insane at Lexington, which he conducted until 1875, when he accepted the Superintendency of the Cincinnati Sanitarium, a private hospital for the insane, suburban to the city, the name of which it bears ; where, in the successful discharge of professional duties, February 11, 1880, he died of structural disease, having nearly completed that term of years beyond which consciousness of existence is the chief compensation of life.

Dr. Chipley was a man of imposing presence and dignified address. His characteristics as a man were loyalty to duty, principle and personal friendships. He was fond of society, especially of men, choosing by preference persons younger than himself. Professionally he was orthodox, adhering to the doctrines and practices of the school from which he received instruction. There was no smell of quackery in his garments. As an alienist he stood well with his co-laborers in the field of psychiatry—among his own people, prominent. His contributions to the literature of science were not voluminous, but always respectable. He was an earnest, intelligent, sincere, practical man and physician—a high-toned patriotic citizen, and under all circumstances a gentleman, in the American acceptance of that designation.

While by his life he illustrated some of the more prominent virtues, claimed as peculiarly " Christian," intellectually he rejected the pretentions of Christian theology, and died as he had lived, a philosopher.

His remains were returned to the earth, and repose in the cemetery at Lexington, Kentucky, surrounded by scenery endeared to him by all the incidents of a happy childhood, and the achievements of an active and successful career.

———

Dr. Joseph T. Webb was born in Chillicothe, Ohio, in 1827. Here he received his preliminary education. He entered the

Ohio Wesleyan University, at Delaware, Ohio, and graduated
with honors in the year 1848. Soon afterward he began the
study of medicine in the office of his father, an eminent physician of Chillicothe, and in due time matriculated at the Transylvania Medical College, Lexington, Kentucky, where he graduated in 1852. He there opened an office and continued the
practice of medicine in the city of Cincinnati until 1858, when
he engaged in the manufacture of varnish in partnership with
John Piaff, and continued in this business until the opening of
the war, in 1861, when he entered the volunteer service of the
United States Army, as Surgeon of the 23d Ohio Volunteer Infantry, in which capacity he served until the close of the war, in
1865. Not long after this date he married Miss Anna Matthews,
and traveled throughout Europe until 1871, when he was elected
Superintendent of Longview Asylum for the Insane. He resigned this office in 1874, on account of ill health, and traveled
again over Europe and America in the hope that change of air
and location might be found of benefit to him. He died at
Minneapolis, Minnesota, April 27, 1880, at the age of fifty-three.
Thus passed away the soldier, the scholar and the gentleman in
the prime of life, in the midst of his family and friends, surrounded by all that wealth, honor and distinction could procure.
He was a man of great executive ability, generous, sympathetic,
impulsive, and in his nature, kind and obliging, a true friend
and a true gentleman. He was brother-in-law of Hon. R. B.
Hayes, ex-President of the United States, and also of Hon.
Stanley Matthews, Associate Judge of the Supreme Court of the
United States.

Dr. Robert F. Baldwin, the eldest son of Dr. A. Stewart Baldwin and Catharine Mackey, was born in Winchester, Frederick
County, Virginia, on the 16th day of August, 1829. As a child
he was exceedingly attractive, possessed of rare personal beauty
and a merry, joyous temper. As a youth, he was vigorous in
health, attentive to his studies, entered with ardor into all the

sports of boyhood, excelled in horse-back riding, and was generous to a fault. Blessed with parents whose great aim was always to make home the most attractive spot, and surrounded by
very favorable circumstances, he early developed those domestic
traits which made him, in after years, so thoroughly to centre
his happiness in home life. After attending the academy in
Winchester for several years, and impressed with the idea that he
should follow in the footsteps of his father and grand-fathers, all
of whom had gained high reputations as physicians, he entered
the office of his father and his uncle, Dr. Robert F. Baldwin, as
a medical student. Subsequently spending the year 1848-9 at
the University of Virginia, thence he went to the University of
Pennsylvania, where he graduated in medicine in 1851. Remaining during the summer in hospital practice in Philadelphia,
he returned to Winchester and commenced the practice of medicine in partnership with his father. He rapidly gained the
confidence of the intelligent and refined community in which he
lived, and in a few years had a well-established practice. He
was so highly esteemed by his professional brethren that they
often called him to their assistance in consultation. In October,
1856, he married Miss Carrie Barton, of Virginia, a lovely bride,
who ever afterwards adorned with excellence, grace and affection, his family and social circle.

At the beginning of the sectional war of 1861, he espoused the
cause of his native State, was commissioned a Colonel of Militia,
and assigned to the 31st Virginia Infantry. While attempting
to check the advance of a greatly superior force under General
Lander, near Bath, in West Virgina, he, with a few of his command, after a gallant resistance, was captured. In this action he
bore himself with such gallantry and cool courage as to excite
the admiration of General Lander, which was expressed in appropriate terms, in after years, by a member of his staff, when
returning his sword, surrendered on that occasion. Remaining
in Camp Chase and Fort Warren until 1862, he was exchanged
and returned to Richmond, where he was commissioned a Surgeon and assigned to duty with the 5th Virginia Infantry, in the
Stonewall brigade. He served with this command for several

months, but owing to some dyspeptic trouble, he was relieved from field duty, ordered to Staunton, and assigned as Surgeon in charge of a general hospital. He discharged his duties in that capacity with great acceptability, and there developed the administrative talents for which he was afterward conspicuous when called to another field of usefulness. Returning to Winchester in 1865, he pursued his practice with the same unselfish devotion until the summer of 1874. For some months previous to that time, the cares and anxieties resulting from the extreme and protracted illness of some members of his family, the loss of his venerated father, and subsequently of his devoted mother, with the added labors of a large practice, enfeebled his constitution. While in that condition, after a long ride on a hot summer day, he was seized with a violent pain in his right eye, from which he suffered intensely for several weeks. After rallying to some extent from his prostration, he went to Baltimore, consulted Dr. Chisholm, who deemed an operation necessary and extirpated the eye, finding a small tumor on the optic nerve. He returned, and in a short time was apparently restored.

A vacancy having occurred in the Western Lunatic Asylum, by the death of its superintendent, Dr. Francis T. Stribling, it devolved upon the Board of Trustees to elect his successor. Dr. Stribling had filled the position with distinguished honor to himself and acceptability to the citizens of the State ; the institution, under his wise and judicious management, extending over a period of many years, had acquired and established such a reputation that the Board realized the importance of selecting one possessed of the qualifications necessary, not only to maintain that degree of excellence which it had acquired under its late superintendent, but to conduct in still more extended fields of usefulness, as the demands increased for its enlargement. After due consideration, the Board decided to confide this important trust to Dr. Baldwin. In accepting it he realized the weighty responsibilities incurred, and the more so that in his professional career he had not made the subject of insanity an object of special study and practice ; but, trusting with a Christian's faith upon the guidance of a kind Providence, with self-reliance acquired from long experience in his profession, a matured judgment and manly nature, he entered upon the duties of his position, giving himself wholly and earnestly to his work. Combining good attainments with fine executive ability, he soon gained the respect and confidence of his associates, and so conducted the institution in all its interests, that his administration received the cordial endorsement of his Board of Directors, endeared him to the inmates of the asylum and their friends, and obtained for him the confidence and esteem of the community as a faithful and efficient public officer.

Most of the members of this Association will remember the deep interest he took in all its proceedings, contributing what he thought might be beneficial to others, and in trying to obtain for himself a knowledge of the most improved methods of asylum management, by which to promote the welfare of his own institution. The last meeting of the Association which he attended, was the one held in Washington city, in 1878. A few weeks previous to the meeting held in Providence, in 1879, the writer received a letter from him, in which he expressed deep regret that he could not be present, said he was suffering with his eye, which had been again operated upon, and that he could not stand to be in a crowd. He was not benefitted by the second operation, and it soon became apparent to his friends that he could survive but a few months. His wife died during the summer, and on the fourteenth of November he was called to rest from his labors and sufferings.

Dr. Baldwin belonged to that class of men not distinguished for any peculiar characteristics, but possessed all those qualities essential to the highest type of manhood. With good natural endowments and liberal culture, he was eminently a practical man. With an inherited fondness for his profession, he pursued it with energy and in the most catholic spirit. While animated with a laudable emulation, he observed the courtesies of the profession with the most scrupulous care and guarded the reputation of a worthy brother of the fraternity as his own. Responding to the calls of the affluent he was compassionate to the afflicted and needy, and like the good Samaritan was ever ready to go to

their relief. As a public officer he held the trusts confided to
him as sacred, and administered them with inflexible fidelity.
Exemplifying in himself the principle of honesty and integrity
upon which his character was based, inspiring others with the
same spirit of zeal and unselfish devotion by which he was actua-
ted, with the capacity to design and the firmness and energy necessary to have his plans executed, he was making, in connection
with the institution to which he was attached, a just and enduring fame, which caused his death to be the more deeply regret-
ted, occurring as it did in the prime of his manhood and in the
midst of his greatest usefulness. This regret was participated in
and touchingly expressed by many of the inmates of the Asylum
who had been the objects of his care and tenderest sympathies.
In all the relations of life, both public and private, his character
shone forth brightly as an honored type of the Christian gentleman.

In 1863, while nursing a brother, a most promising young
physician, in his last illness, he determined to become a follower
of Christ, and soon after united with the Protestant Episcopal
Church in which he was for many years a vestryman. His walk
was that of an humble and sincere Christian. A dutiful son, a
cherished brother, a devoted and affectionate husband and father,
and genial and social in disposition, he was the charm of the
family circle in which he found his chief happiness. His doubly
bereaved children will delight to cherish his memory, not only
as a prominent and honored citizen of the State, but more, as a
kind and affectionate father who never chided but in gentleness
and love, and who delighted to make their youthful days bright
and joyous.

Intellectual and cultivated, manly and true, brave and generous, firm and energetic, frank, amiable and gentle and adorned
by the Christian graces, he combined the virtues and excellencies
of a character of remarkable symmetry, and which was most fitly
and tersely expressed by the Board of Directors of the Asylum
in their annual report succeeding his death, in the following
words : " A man of ability and administrative tact, he united
to the highest factors of a true manhood, the gentleness and

graces of a woman, rounded out into the highest type of the
Christian gentleman;'' and still further by Governor Holliday
when communicating the fact in his annual message to the General Assembly of Virginia :   " It is also my duty to inform you
of the more recent death of Dr. Robert F. Baldwin, Superintendent of the Western Lunatic Asylum.   When elected a few
years ago he was an eminent physician in the full practice of his
profession.   He entered upon the discharge of his duties at the
Asylum with a high sense of the responsibilities which belonged
to the office, and so bore himself as very soon to win the confidence of all by his ability, faithfulness and diligence in the discharge of the great trust.   His death is a loss to the institution
and to the profession of which he was so honored a member.''

His remains rest peacefully at a spot of his own selection in
the beautiful valley which was the scene of his labors.   His memory will long remain green in the hearts of the Virginia people,
and may well be cherished by this Association as one, who in
the few years of his membership, devoted the best energies and
impulses of his nature to aid in its chief object,—the amelioration of the condition of the insane.

Wm. Maclay Awl was born in Harrisburg, Penn'a, on May 24,
1799, his mother having been a lineal descendant of John Har-
ris who founded Harrisburg, and the daughter of Wm. Maclay,
the first Senator of the United States from Pennsylvania; and
while quite young the family moved to a farm at a short distance
from Sunbury, Penn'a.   When fifteen years of age he was sent
to the Academy in Northumberland kept by Rev. Isaac Greer
and after his death by his son Robert C. Greer, afterwards Judge
of the Supreme Court of the United States, and there he acquired
all his preliminary education.   He studied medicine with Dr.
Samuel Agnew, of Harrisburg.   He attended one course of lectures in the session of 1819-20 in the University of Pennsylvania, which seems to have been the only course he attended.

though he received the honorary degree of M. D. at a later date from Jefferson College. In the spring of 1826 he started on foot for Ohio and settled first at Lancaster, and an important surgical operation there performed gave him his first introduction into practice. After moving several times from place to place he finally settled in Columbus, Ohio, in 1833. His attention was first called to the care of the insane by a case which occurred while in Somerset, Ohio, which he was called upon to treat, a case of violent acute mania. Shortly after his settlement in Columbus an epidemic of Cholera occurred which gave him abundant opportunities of practice in the community at large and also in the Penitentiary.

On Jan. 5, 1835, he attended a convention of medical men of Ohio which had been called by himself and several others to take some measures towards the care of the insane and the education of the blind. A memorial was presented to the Legislature on these subjects, and an appropriation was obtained towards the erection of a hospital for the insane, and Dr. Awl was appointed one of the Trustees to build it. He in company with two others visited the Eastern and Middle States to gain information on the subject. The building was completed in 1838 and Dr. Awl resigned as Trustee and was appointed Superintendent.

He was one of the originators of the Ohio Institution for the Blind and was always deeply interested in that Institution and was physician of it at the time of his death. He continued in charge of the Lunatic Asylum until 1850 when he was displaced by that system of political appointment which has so unfortunately prevailed in Ohio from that day to this.

Dr. Awl was one of the original thirteen members of this Association and always manifested a warm interest in all its proceedings. He was Vice President from 1846 to 1848 and President from 1848 to 1851. He was also one of the original members of the American Medical Association, of which he was one of the first Vice Presidents. "In 1861 he was appointed by Gov. Denison one of the Board of Medical Examiners for surgeons of the Ohio Regiments and was President of that Board during its ex-

istence. In 1862 Gov. Tod appointed him Superintendent of
the State Capitol, which office he held for six years. In 1873
Gov. Allen appointed him Physician to the Blind Asylum, which
office he held to the close of his life and prepared his last report
only a few days before his death." He attended the meeting of
the Association in Philadelphia in 1876.

For several years he had been suffering from a complication of
disorders and passed away quietly on November 19, 1876. He
had for many years been a member and elder in the Presbyterian
Church. Dr. Awl was of a cheerful, lively disposition with a
great fund of natural genial humor which made him a very pleasant companion and united with great tact and sound common
sense, served him admirably in dealing with the insane.

———

Dr. Clement Adams Walker was born in Fryeburg, Maine,
July 3, 1820. He died suddenly after several years serious illness
April 26, 1883, being 62 years and 9 months of age. His boyhood was passed near the White Mountains of New Hampshire
and almost in the shadow of Mt. Kearsage. The beautiful Saco
intervale and Jockey Cap over-looking Lovewell's Pond, often
recalled to him the stirring traditions of Indian warfare.

He fitted for college at the Fryeburg Academy, a school once
honored by the instructions of Daniel Webster, and still flour-
ishing. He graduated at Dartmouth College in the somewhat
remarkable class of 1842, of which he was not the least distinguished member. Among his classmates and college-mates I
now recall Hon. L. F. Brigham, late Chief Justice of the Superior Court of Massachusetts; Hon. Isaac Ames, late Judge of
Probate for Suffolk County; Rev. Dr. Samuel J. Spalding, of
Newburyport; Dr. J. Baxter Upham, of Boston, and Dr. John
E. Tyler, for many years Superintendent of the McLean Asylum
at Somerville, and an honored member of this Association. He
enjoyed the life-long friendship of all these and many others of
his class. His intimacy with Dr. Tyler began in College and

continued with more than brotherly affection until the death of the latter, a few years ago. His power of making and keeping friends was one of the strongest points of his character.

During his college career his health gave way and he traveled in the South, teaching school in Virginia and making some valuable acquaintances there. He had suffered from hemorrhage from the lungs, which led his friends to fear a fatal result. He afterwards acquired an apparently vigorous physique which was severely tested by his thirty years of active hospital life. He was a little above the medium height and became stout in middle life. His eyes were dark and piercing, his lips expressive of firmness, the nose large and his hair straight and jet black in youth, but turning at thirty-five to white, with his snowy beard gave him the aspect of a vigorous old age in early manhood. He graduated in medicine at Harvard University in 1850, and began practice at South Boston under Dr. Charles H. Stedman, who was then physician to all the city institutions located there, including the Boston Lunatic Hospital. In 1847–9, when cholera and ship-fever were prevalent among the poor emigrants at the quarantine station at Deer Island, he volunteered with his classmate, Dr. Upham, to assist in the fever-sheds and rude hospitals erected there for temporary use. He entered on the work of managing these unfamiliar and dreaded diseases with characteristic promptness, courage and skill. Dr. Upham's reputation was speedily established by an able monograph on ship-fever; and Dr. Walker's no less so by his success in dealing with the intractable diseases above mentioned. July 1, 1851, Dr. Walker was appointed Superintendent of the Boston Lunatic Hospital, which position he held until his resignation on account of ill-health, January 1, 1881, a period of nearly thirty years.

This hospital, built in 1839, had been in charge of Dr. Butler, its first superintendent, and Dr. Stedman, whom Dr. Walker succeeded, a period of twelve years. In its rear was a semi-detached building known as the "Cottage," fitted up with cells like those of a police station, for the violent insane. Such cells were supposed to be a necessary adjunct to an hospital for the insane in those days. Dr. Walker at once advised their disuse, and, in a short time, succeeded in having them abandoned by gradually placing their occupants into the wards of the main building. He thus became the pioneer in the discontinuance of cells in the treatment of the insane in this country. He was remarkable for bringing things to pass. Whatever he took in hand he gave his whole mind to : and his clear intelligence, strong will, and skillful management accomplished many things seemingly impossible. In the care of the insane, these qualities gave him a great advantage over obstacles, and exerted a powerful moral influence upon patients and their friends. He never knew when to give up a case. With death at the very door he persisted in active and sometimes successful treatment.. While not neglecting judicious alimentation he had more faith in medicines than is fashionable at present. While life lasted there was not only hope, but active help for all his patients. In many ways he improved his hospital, elevated the standard of treatment, diminished restraint, and brought about needed changes and reforms. For many years his advice was implicitly relied on by successive Boards of Visitors and Directors.

He early recognized the necessity for better accommodations for the city's insane, and for years labored earnestly for this object, until success nearly crowned his efforts. A site for a new hospital was purchased, plans made and adopted, and an appropriation passed only to be vetoed by the Mayor, who opposed the project. It was said that the site was exposed, remote and difficult of access. But the substitute hospital at Danvers is as much exposed, ten times as remote and far more difficult of access. The site at Winthrop, said to be uninhabitable, is surrounded by dwellings, newly erected, is reached hourly by rail, and has just been sold for three times its cost to the city. This veto was a severe blow to his hopes, and he had only the sad satisfaction of seeing the city's plan of construction adopted at Danvers and of having the medical supervision of the work in behalf of the Commission which had it in charge.

As an expert in mental disease, Dr. Walker was frequently called in Court in his own and other states. His opinions being deliberately formed and clearly expressed, carried weight in consequence. His written opinions, reports and medical papers were always carefully prepared, condensed in expression and logical in method. His hand-writing even expressed his character, in its peculiar squareness and solidity. In dealing with men, a rare combination of strength of mind, sound judgment, tact and well-chosen language gave him great influence and made him a a safe adviser, a useful advocate and friend. He made the most humble, whose cause he espoused, feel that his chief desire for the moment was to serve his interests. The patience with which he entered into the details of another's troubles, or listened to the tedious recital of symptoms, was only equalled by the persistency with which he devoted himself to their relief. He left no stone unturned to accomplish his benevolent purposes. He was large-hearted, sympathetic and generous to a fault, and now and then was made the prey of ingenious schemers through an excess of misdirected sympathy. His social feelings were strong and his acquaintance grew in many directions. He was prominent in the Masonic order, reaching the highest degree attainable in a very short period, and devoting much time and energy to the subject while his interest lasted. He was an active member of this Association from 1851 until a short time before his death, and was your President for three years preceding his resignation of that office in 1882. When in good health he was usually present and took a leading part in your deliberations. During the war he was appointed Inspector of Hospitals and made a tour of service in the West. In 1872 he made a brief visit to Europe. A few years since, by the influence of the German Consul, he was presented with the decoration of an order of nobility for his humane treatment of an insane German citizen in Boston. He was a member of numerous medical societies, a complete list of which cannot be given at present.

Dr. Walker was buried with Masonic honors and his funeral was attended by many of those whose physician, friend, or benefactor he had been. Many a depressed and despairing sufferer whose burden he had lightened or removed has reason to bless his memory and to mourn his loss. You may perhaps ask if there was no defect or weakness in this excellent character I have attempted to describe. I should reply that there were many, but they were the defects of an exceptionally strong and noble nature. The world is full of minds made up of weaknesses in every possible combination and we need no such examples. Here was a man of positive qualities, of great natural strength and excellence, whose influence was remarkable upon all with whom he came in contact. Let us then emulate the good and forget the weakness that was in him, since we may soon need a like charity for our own numerous failings.

I cannot better close this sketch than by adding the following words by the Rev. Edward Everett Hale : " He was the personal friend of every patient, and brought to the miracle of cure the only power which can effect it,—the loving sympathy of the physician. He fairly commanded his broken patients, in instances too many to name, by what we choose to call the magnetic power of his personal care. They believed in him. They did what he bade them. Behind all the resources of medicine and treatment, he had this requisite of victory, that he made them believe they would get well. Thirty years of such life exhausted him completely. We wonder that he lived so long. You cannot give out forever. Two years since he retired from the charge of the hospital, and, after a period of rest, which did fully restore him, he entered into private practice at the South end. But the end had really come. Symptoms of disease again and again alarmed his friends though nothing would alarm him ; and now, too soon for them, they have to deplore his sudden death. One looks at it as at the loss of a soldier who is shot down in battle at the head of a column. Any one who remembers, as the writer of these lines does, the homes this man has made happy, the lives he has restored to duty and joy ; and who knows that in working such cure, his will, his resolution and determination were eating away even in the power of life by which he wrought them, feels that here is, indeed, one instance more of the way in which a brave man is willing to die for mankind."

Dr. Isaac Ray was so intimately known to the members of this Association, and was so universally honored for his great ability and the general soundness of his views on all subjects on which he wrote, that little more can be said in the necessarily limited space to which a notice of him, for insertion in the proceedings of the Association, must be confined. The death of Dr. Ray leaves but three of the original members of this body—two with their armor still on, and in active service, while the third enjoys, in his retirement, the honors due to long and faithful labor in this field of benevolent usefulness.

Dr. Ray was one of the " original thirteen " Superintendents who established " The Association of Medical Superintendents of American Institutions for the Insane" in 1844 ; was its Presi-, dent from 1855 to 1859, and always took a very marked interest in its proceedings. His papers read at its meetings were numerous and of great ability. Many of these cannot fail to take a permanent place in the literature of the profession. Dr. Ray formed the highest estimate of the importance of this Association and of the value of the work which it had done, and especially believed that to it the insane were to look for most of the changes which were likely to be made in their care and management, which could lay any claim to be for their best interests or really worthy of the name of progress. Its " propositions," now more than a quarter of a century old, and having stood the test of trial in every section of the country, had in him a staunch defender, and his practical knowledge and extended observation of other systems gave to his views an especial worth. At the meeting of this Association in Providence, in 1879, he was the recipient of distinguished honors from his old associates and friends from the Rhode Island State Medical Society, of which he had been President, and from Brown University, which, on this occasion conferred on him the honorary degree of Doctor of Law. Of the principles established by the Association, Dr. Ray, like all his most experienced brethren, was particularly decided in regard to the importance of a proper organization. He knew, from extended observation of other schemes, that only a single head, controlling, as he must have the responsibility of all departments, can be relied on for a permanently successful administration, and he lost no suitable opportunity for enunciating this most important principle, no departure from which he believed could ever be justified as tending to promote the best interests of the insane.

The subject of this memoir became a fellow of the College of Physicians of Philadelphia in July, 1868 ; he was always interested in its proceedings, often reading valuable original papers, and generally participating in its discussions. The estimate in which he was held by his associates was shown by the action of the College on the occasion of his death, and by the resolutions subsequently adopted and ordered to be entered on its minutes. It may safely be said that few men, at home or abroad, have attained a higher eminence as members of the medical profession, as directors of institutions for the treatment of the insane, and as writers on insanity and medical jurisprudence, than Dr. Ray. He became a resident of Philadelphia in the autumn of 1867, and from that time took an active interest in whatever tended to advance the welfare and prosperity of his adopted home. He was a frequent contributor to the daily press, and almost all the subjects that were generally discussed, in one way or another, had the benefit of his mature judgment and thoughtful consideration. He was always ready to give his time to the promotion of objects of benevolence, and to render assistance to those who were specially unfortunate, and his very extended and varied experience secured for his opinions a more than ordinary degree of respect and public confidence.

Dr. Isaac Ray was a native of Massachusetts. Born of highly respectable parents, in the town of Beverly, on the 16th of January, 1807, he there commenced his earliest education, subsequently entering Phillips Academy at Andover, and afterwards Bowdoin College, where he remained till compelled by ill-health to leave his studies, which he had been prosecuting with great assiduity. As soon as his health was sufficiently restored, he began the study of medicine in the office of Dr. Hart, of Beverly, completing his studies under Dr. Shattuck, a distinguished physician in Boston, and ultimately graduating at the Medical De-

partment of Harvard University, in 1827. .In that year he began
the practice of his profession at Portland, Maine, and while there
he delivered his first course of lectures on botany—a branch of
science for which he had a great fondness. It was at one of
these lectures that he first met the lady whom he subsequently
married in 1831, Miss Abigail May Frothingham, a daughter of
the late Judge Frothingham of Portland, who still survives him
and with whom he lived in a most happy union for a period of
just two months less than fifty years. From this marriage two
children were born—a daughter, with rare traits of loveliness,
who died at the age of fourteen, and a son, to whom further
allusion will be made in a later part of this notice.

About two years after Dr. Ray had commenced the practice of
medicine in Portland, Maine, inducements were offered to him
to leave that city and settle in Eastport, in the same state ; there.
soon after, he fixed, as he then supposed, his permanent resi-
den ce.

It was at this time, while living in Eastport, that Dr. Ray first
had his interest excited on the subject of insanity and the treatment of the insane, and especially in reference to matters connected with the branch of medical jurisprudence relating to it.
The prevalent views on all these subjects were then far behind
what are common at the present day, and led Dr. Ray to prepare
a work, "The Jurisprudence of Insanity," since generally recognized as one of the highest authorities in this department of
medico-legal knowledge, and quoted alike by alienists, lawyers
and all others interested in the subject, at home and abroad.
No better evidence of its being generally appreciated need be
given than the fact that six editions have been exhausted in this
country, while it was a source of grief to Dr. Ray that his condition of health rendered it impossible for him to prepare a
seventh, which had been asked for by his publishers, and for
which he had on hand interesting and important materials. The
steady increase of popularity attained by the " Jurisprudence of
Insanity," as might have been anticipated, led to a change in
the tone of the criticisms made in regard to it. From being

originally adverse in many quarters, they became highly com-
mendatory everywhere.

Dr. Ray was appointed Medical Superintendent of the State
Hospital for the Insane, at Augusta, Maine, in the year 1841, and
this led to his permanent removal from Eastport. He immediately assumed the duties of this position, residing in the institu-
tution till he was invited by the Board of Trustees of the Butler
Hospital at Providence, Rhode Island—which was then about to
be organized—to become its Superintendent.

The experience of hospital life and management in a State
institution was of great importance to Dr. Ray. It enabled him
to detect and expose many of the weak points to which this class
of hospitals is made liable, and gave to the emphatic views which
he afterwards expressed, a particular value, from the practical
nature of the observations which had led to them. He never
failed to censure in the strongest terms the evil results of a political management, of giving to the directors of such institutions
a personal and pecuniary interest in their business affairs, and of
confiding to those in no way qualified by education and experience the control of the important matters of treatment and
government assigned to them. He was always ready to denounce
an institution without a head, as much as one with many heads,
as a monstrosity that could not, unless under extraordinary circumstances, be more than a temporary success ; and his enuncia-
tion of sound views on all such subjects, on all proper occasions,
has exercised an important influence in every part of the country.

Dr. Ray found his position at Providence a specially pleasant
one. His labor was much less arduous than it had previously
been ; he was enabled to carry out his own well-considered plans,
and it afforded him a long-desired opportunity to visit many of
the more prominent institutions for the insane in Great Britain
and on the Continent. The fruits of abundant practical knowledge and a careful study of the whole subject, gave him special
qualifications in these investigations to detect errors, and to
make a trustworthy comparison of the actual advantages and disadvantages to be found in the institutions at home and abroad.

With this view Dr. Ray sailed for Europe soon after his appointment, and in this manner passed the summer months of 1845. He spent the next two years in superintending the erection of the Butler Hospital, which was opened for the *reception* of patients in 1847. Then taking up his residence in the hospital, he remained there superintending its affairs with great ability, and to the satisfaction of all who were in any way connected with it, till January, 1867, when his impaired health compelled him to resign this position to which he was so much attached, and in which he had done so much to elevate the standard of hospital treatment for the insane. This relief from labor and from all the cares and anxieties unavoidably incident to the conscientious superintendence of a hospital of this description made him greatly enjoy a rest, such as he had never before taken except during the trip to Europe. He spent most of the year in visiting his professional brethren in different parts of the country, and in selecting a place for his permanent residence, finally accepting the city of Philadelphia. Here he continued to live at his residence on Baring Street till his death on the morning of the 31st of March, 1881, being then in the seventy-fifth year of his age.

The change from a New England climate to that of Philadelphia, and the rest from constant labor which was permitted him, made a great improvement in Dr. Ray's health. He increased his literary work, enjoyed engaging in matters of general public interest, and found himself able to take an amount of physical exercise to which of late he had been a stranger. His regained strength enabled him also to accept calls in consultation from his professional brethren, and especially as an expert in legal and criminal cases in which his services were frequently solicited. Dr. Ray was a member of many professional and scientific associations. Wherever he was thus associated he was noted for his active interest, and for the part he took in the preparation of papers, and his participation in any discussions that might take place. Dr. Ray was one of the founders of the Social Science Association, and was always an intelligent student of every subject which came under its consideration. His papers read before it, and his views in all matters that received its attention, were distinguished for practical good sense and advanced conclusions in regard to the welfare of the community.

He was, at one time, a most useful member of the Board of Guardians of Philadelphia, giving his valuable time to the duties of the post, which, conscientiously performed, could not fail to be onerous. His experience and his devotion to the insane led him to take an active part in the work of that department. He was not long in detecting its grave defects, and in suggesting the proper remedies; but the minority, with which he acted, had the power to introduce but a few of the reforms which they knew to be indispensable. It is one of the remarkable events of the times that the public authorities were willing to dispense with the unremunerated services of such a man as Dr. Ray, to make a place for some one who had not, and who from his previous life could not have, the first element of knowledge fitting him for a post, one of the most important duties of which was to secure for the insane a liberal and enlightened treatment.

Dr. Ray delivered two courses of lectures on " Insanity and Medical Jurisprudence " before the class of one of the medical colleges of Philadelphia, but, as usually happened in regard to his public labors of the kind, they were without compensation, and demanded an amount of time which he could ill afford to continue to give. While it must be acknowledged that it is not easy to give such a course of instruction to students as Dr. Ray was competent to impart, still it must be conceded that lectures of this kind in every medical school would do much to advance the study of mental diseases and their treatment; would make the profession, and through it the public, better able to detect the many defective schemes of organization now presented for hospitals for the insane, and would lead to a much higher order of discussions in many of the meetings held ostensibly for the special improvement of the care of the insane, and in others in which their management of late has seemed to be the favorite subject for consideration.

Dr. Ray was about the medium stature, but did not possess a very robust constitution. His features were marked and his gen-

eral expression grave. He had an abundance of rather stiff hair,
which of late years was entirely white, and from his way of treat-
ing it, it was commonly somewhat in disorder. His manner was
dignified, his language clear and distinct, and in speaking or
writing he always used a pure English, and attracted the attention of his auditors no less by his personal appearance than by
his manner of delivery and the matter of his remarks.

Dr. Ray, for many years, had been troubled with a chronic
cough which seemed to be bronchial in its character. Although
annoying, this cough did not appear materially to affect his general health, and after taking counsel from the most able of his
medical brethren, he seemed to have concluded that his malady
was one not likely to be removed by treatment. The great
change in Dr. Ray's health, which occurred in the latter part of
1879, was evidently more the result of a great and unexpected
family affliction, than of his previous condition. His only son,
Dr. B. Lincoln Ray, was a highly educated physician, living
with his parents in Philadelphia, greatly valued by them, and distinguished as a student and writer of very marked ability. Of
vigorous personal appearance, he nevertheless was conscious for
some time before his death, of an impending cerebral malady.
which gave him very serious apprehensions. On the evening of
the 7th of December, 1879, these indications of brain trouble
were suddenly developed into an acute attack, which, with great
suffering, ended his life in the short period of forty-two hours.
To this only son, his parents had looked forward as a comfort
and support in their advancing age. They had been proud of
his abilities, and from his filial devotion, and as an appreciative
exponent of his views, his honored father had hoped to have jus-
tic done to his labors, as an author, by one abundantly capable
to give a proper exposure to those who had not hesitated to use
his thoughts and occasionally his very language, while forgetting
to give the slightest word of acknowledgment. It was not wonderful that the sudden death of this son, at the meridian of life.
should have left results of no ordinary character. This sad
event, so unlooked for, was a shock to the father, which did more
to prostrate his health and strength than would have been done

by years of customary labor. With his intimate friends he was still the same genial character, still interested in whatever concerned his profession or his fellow man ; but he ceased to write, complained of what had formerly been a pleasure now becoming a toil to him ; found his flesh wasting and his strength diminishing, and frequently showed a sadness quite unnatural to him. Gradually he became less and less able to take his usual amount of out-door exercise, or to attend to the calls of professional business. From the early part of December, 1880, he remained in his house, still seeing his friends, interested in his books and in what was going on in the world and in his specialty, but steadily losing weight and strength. To avoid the fatigue of going up stairs, he ultimately made his library, in the second story, his lodging room, and spent the greater part of every day at his front window in the adjoining apartment, reclining in an easy chair— a highly valued present, years before, from a beloved professional brother—and looking out from it upon what was passing on the active thoroughfare before him, and on the beautiful gardens of the houses opposite his residence. From his daily increasing weakness his friends realized that the end must be near. On the evening of the 31st of March, 1881, he retired at about the usual hour. After being in bed he had one troublesome spell of coughing, but then slept quietly, only once in the early morning, inquiring the hour. So peacefully did he rest, and so calm was his sleep, that he made no sound of any kind, nor moved a muscle, as far as could be heard ; and when approached somewhat later, there had been no change in his position, but life had departed, and only what was mortal remained of this noble and useful man.

Dr. Ray was a man of great versatility of talent. His ability as a writer is well known, and his conversational powers were remarkable. He had a great facility in adapting himself to any society in which he might be placed, and was equally agreeable to the grave professional man, or to the specialist, as to those of tenderest age, with whom he was usually a great favorite. While to a stranger Dr. Ray's manner might at first appear somewhat austere, this impression was removed by a very limited intercourse. By his intimate friends and associates he was specially honored and esteemed and no one was more cordially welcomed in the social circle.

A list of Dr. Ray's writings, which has been preserved, shows how industrious an author he was, and how multifarious were the subjects in which he took an interest. From 1828, when his first publication of which any record has been kept, was made, down to 1880, during which year he published his last contributions in the press, it will be seen that but a single year passed in which something original was not noted.

———

Dr. R. H. Gale was born in Owen County, Kentucky, on the twenty-fifth day of January, 1828. His life, though cut off a little past middle age, was singularly eventful. Graduating when quite young from Transylvania University at Lexington in his native state, he entered the office of his father, an eminent and popular physician, as well as a wealthy and influential man, and after the usual term of pupilage, under the care of so interested and capable a preceptor, he was enrolled in the classes of the Jefferson Medical College of 1847 and 48, graduating with excellent standing the latter year. His first location in the pursuit of his profession was at Covington, Kentucky, where it is said his practice was signalized from the beginning by marked success. While in this field he became a staff officer of the Cincinnati Commercial Hospital. After very creditable public service and while possessed of a flattering and remunerative private clientage, he was induced by his family and their friends to change his location to the midst of the community in which he had been reared : where his personal worth was appreciated, it might be said, to a degree of partiality, and his professional capability and skill were recognized at once. A man of lively sympathies and of a generous and genial nature, he could never feel indifferent as to whatever affected in any way those among whom he lived and moved. He was distinctly and distinctively one of the people. Influenced by their wishes, he was twice elevated by their suffrages to the office of County and Probate Judge of Owen County. Subsequently he served his county one or more terms in the legislative councils of the State, assuming a prominent part in their proceedings and leaving a highly creditable and flattering record. At the beginning of the war, impelled by his ardent sympathies with the South, he entered the service of the Confederate States in Col. D. Harvard Smith's regiment, which constituted a portion of Gen. John H. Morgan's famous command. His health failing from the energetic performance of his very arduous duties, he was obliged to resign his position and quit the service. After the war, he settled in Louisville where he immediately realized the eminence which he had already achieved. He very soon commanded a lucrative practice and assumed a prominent place upon the staff of the City Hospital, where, his tastes affecting surgery most, he took an enviable stand among the many powerful and eminent men then and still identified with the specialty in that institution. Besides devoting considerable time to clinical teaching in the hospital, he also gave lectures for several seasons in the Louisville Medical College. He was chosen about the same time Secretary, who was also ex-officio financial manager of the Physician's Medical Aid Society. In 1873, he was appointed Surgeon to the Louisville, Cincinnati and Lexington Railroad, and a year later by the Paducah road to a similar position. In this capacity he served these roads till 1879, when he was appointed by Gov. Blackburn as Superintendent of the Central Kentucky Lunatic Asylum, in which position he continued till the day of his death which occurred, as remarkable, on the day fixed for his resignation of the office to take effect. Lately, Dr. Gale had realized, very sensibly and painfully, the aptness of the pithy and pointed words of somebody, that "a superintendent of an asylum for the insane dwells ever upon a volcano liable at any moment to erupt a catastrophe." He was both confiding and indulgent and trusted his subordinates perhaps unduly. Unfortunate occurrences, concealed from him, led to charges which challenged investigation and which eventuated in confirmation. Although the great mass of the testimony in the premises went very far to exculpate Dr. Gale himself, and to establish the goodness of his nature and efficiency of his management, still the worry and anxiety incidental to the proceedings so preyed upon his sensitive feelings and already failing health, as, no doubt, to hasten his death.

In 1846, when in his nineteenth year, Dr. Gale was married to Miss M. C. Green, a most charming and estimable lady, whose death in 1880 preceded his own. As the fruit of this union three children survive their parents, one son and two daughters, all married. Only a few weeks ago he was joined in a second marriage ; this time to Mrs. Susan Bryant, an amiable and excellent lady, the daughter of Dr. Hughes, a gentleman of fine fortune and great influence, residing near Springfield, Kentucky. In his personality Dr. Gale was a man physically of an exceptionally fine order ; of commanding size he was well proportioned, gainly and graceful. Socially he was genial and unreserved, while he excelled as an agreeable and entertaining conversationalist. Although possessed of mental endowments and culture much above the ordinary plane, still his breeding and native modesty would never allow these qualities to even seem obtrusive. He died at the residence of his son-in-law, Mr. J. C. Reid, in Owen County, near the place of his birth, on the 22d day of April, ult., in the fifty-seventh year of his age.

List of Officers of the Association.

PRESIDENTS.

Samuel B. Woodward, M. D................1844 to 1848.
Wm. M. Awl, M. D............ ....1848 to 1851.
Luther V. Bell, M. D............ ....1851 to 1855.
Isaac Ray, M. D............................1855 to 1859.
Andrew McFarland, M. D....................1859 to 1862.
Thomas S. Kirkbride, M. D.................1862 to 1870.
John S. Butler, M. D.......................1870 to 1873.
Charles H. Nichols, M. D...................1873 to 1879.
Clement A. Walker, M. D...................1879 to 1882.
John H. Callender, M. D....................1882 to 1883.
John P. Gray, M. D........................1883 to 1884.
Pliny Earle, M. D..........................1884 to 1885.

VICE PRESIDENTS.

Samuel White, M. D........................1844 to 1846.
Wm. M. Awl, M. D.........................1846 to 1848.
Amariah Brigham, M. D....................1848 to 1850.
Luther V. Bell, M. D.......................1850 to 1851.
Isaac Ray, M. D...........................1851 to 1855.
Thomas S. Kirkbride, M. D............ ......1855 to 1862.
John S. Butler, M. D............... ....1862 to 1870.
Charles H. Nichols, M. D...................1870 to 1873.
Clement A. Walker, M. D...................1873 to 1879.
John H. Callender, M. D....................1879 to 1882.
John P. Gray, M. D........................1882 to 1883.
Pliny Earle, M. D..........................1883 to 1884.
O. Everts, M. D ....1884 to 1885.

SECRETARIES.

Thomas S. Kirkbride, M. D.................1844 to 1851.
Horace A. Buttolph, M. D...................1851 to 1854.
Charles H. Nichols, M. D...................1854 to 1858.
John Curwen, M. D........................1858 to

TREASURERS.

Thomas S. Kirkbride, M. D.................1844 to 1855.
John S. Butler, M. D......................1855 to 1862.
O. M. Langdon, M. D.....................,...1862 to 1870.

In 1870, the offices of Secretary and Treasurer were united and John Curwen, M. D., was chosen Treasurer in that year.

Meetings of the Association.

———

The following statement gives the time, place and the number attending each meeting:

1. Philadelphia............October 16, 1844...........13.
2. Washington ....May 11, 1846 ....21.
3. New York..............May 8, 1848 ....20.
4. Utica ....May 21, 1849 ....17.
5. Boston ....June 18, 1850.............28.
6. Philadelphia ....May 19, 1851 ....22.
7. New York..............May 18, 1852 ....26.
8. Baltimore.............May 10, 1853 ....20.
9. Washington............May 9, 1854 ....22.
10. Boston ....May 22, 1855 ....26.
11. Cincinnati............May 19, 1856 ....28.
12. New York.............May 19, 1857 ....35.
13. Quebec...............June 8, 1858 ....24.
14. Lexington, Ky.........May 17, 1859 ....17.
15. Philadelphia...........May 28, 1860 ....34.
16. Providence, R. I.......June 10, 1862 ....20.
17. New York.............May 19, 1863 ....25.
18. Washington...........May 10, 1864 ....20.
19. Pittsburgh............June 13, 1865 ....19.
20. Washington...........Apr. 24, 1866 ....27.
21. Philadelphia...........May 21, 1867 ....32.
22. Boston ....June 2, 1868 ....32.
23. Staunton.............June 15, 1869 ....25.
24. Hartford.............June 15, 1870 ....39.
25. Toronto ....June 6. 1871 ....37.
26. Madison.............May 28. 1872 ....41.

178

List of the Hospitals for the Insane

In the United States and the British Provinces, with the names of their Superintendents and the dates of their terms of service, so far as they could be ascertained.

———

MAINE.

HOSPITAL FOR THE INSANE, AUGUSTA—OCTOBER 14, 1840.

Dr. Cyrus Knapp took charge October 14, 1840; resigned April 14, 1841.

Dr. Chauncey Booth acted as Superintendent until August 12, 1841.

Dr. Isaac Ray appointed August 12, 1841; resigned March 19, 1845.

Dr. James Bates appointed March 19, 1845; resigned February 1, 1851.

Dr. Henry M. Harlow appointed June 17, 1852, having been acting Superintendent after Dr. Bates' resignation; resigned June 1, 1883.

Dr. Bigelow T. Sanborn appointed June 1, 1883.

———

NEW HAMPSHIRE.

INSANE ASYLUM, CONCORD—OCTOBER 28, 1842.

Dr. George Chandler appointed March 6, 1842; resigned ——. 1845.

Dr. A. McFarland appointed August 26, 1845; resigned July 30, 1852.

Dr. John E. Tyler appointed October 5, 1852 ; resigned April
15, 1857.

Dr. J. P. Bancroft appointed May 7, 1857 ; resigned March 31,
1882.

Dr. C. P. Bancroft appointed April 1, 1882.

———

VERMONT.

ASYLUM FOR THE INSANE, BRATTLEBORO'—OPENED DECEMBER 12, 1836.

Dr. William H. Rockwell appointed June 28, 1836 ; resigned
August 19, 1872.

Dr. William H. Rockwell, Jr., appointed August 19, 1872 ; resigned December 11, 1872.

Dr. Joseph Draper appointed December 11, 1872.

———

MASSACHUSETTS.

McLEAN ASYLUM, SOMERVILLE—1818.

Dr. Rufus Wyman appointed March 23, 1818 ; resigned May 1,
1835.

Dr. Thomas G. Lee appointed January 16, 1835 ; died October
29, 1836.

Dr. Luther V. Bell appointed December 11, 1836 ; resigned
March 16, 1856.

Dr. Chauncey Booth appointed March 16, 1856 ; died January
12, 1858.

Dr. John E. Tyler appointed February 12, 1858 ; resigned March
3, 1871.

Dr. J. H. Whittemore in charge to July 1, 1871.

Dr. Isaac Ray from July 1 to October 1. 1871.

Dr. George F. Jelly appointed October 13, 1871 ; resigned June 1, 1879.

Dr. Edward Cowles appointed June 1, 1879, but entered on the duties December 11, 1879,

Dr. Frank W. Page acting as Supt. in the interim.

BOSTON LUNATIC HOSPITAL, SOUTH BOSTON—1839.

Dr. John S. Butler appointed September 16, 1839 ; resigned October 10, 1842.

Dr. Charles H. Stedman appointed October 10, 1842 ; resigned July 1, 1851.

Dr. Clement A. Walker appointed July 1, 1851 ; resigned January 1, 1881.

Dr. Theodore W. Fisher appointed January 1, 1881.

STATE LUNATIC HOSPITAL, WORCESTER.

Dr. Samuel B. Woodward appointed September 26, 1832 ; resigned June 30, 1846.

Dr. George Chandler appointed July 1, 1846 ; resigned April 1, 1856.

Dr. Merrick Bemis appointed April 1, 1856 ; resigned May 31. 1872.

Dr. B. D. Eastman appointed July 5, 1872 ; resigned March 1, 1879.

Dr. John G. Park appointed March 1, 1879.

A new building was erected some distance out the city of Worcester and the old hospital building was retained as an Asylum for the Chronic Insane and was opened on October 23, 1877 and Dr. John G. Park was appointed Superintendent October 1. 1877, and remained until March 1, 1879.

Dr. Hosea M. Quinby appointed March 1. 1879.

STATE LUNATIC HOSPITAL, TAUNTON.

Dr. George C. S. Choate appointed October 24, 1853 ; resigned December 7, 1869.

Dr. W. W. Godding appointed April 8, 1870 ; resigned September 1, 1877.

Dr. J. P. Brown appointed January 4, 1878.

State Lunatic Hospital, Northampton.

Dr. Wm. H. Prince appointed August 20, 1857 ; entered on duty October 1, 1857 ; resigned April 1, 1864.

Dr. Pliny Earle appointed July 2, 1864.

State Lunatic Hospital, Danvers.

Dr. C. S. May appointed January 12, 1878 ; resigned August 9, 1880.

Dr. Wm. B. Goldsmith appointed February 17, 1881.

RHODE ISLAND.

Butler Hospital for the Insane, Providence.

Dr. Isaac Ray appointed in 1845, service to commence May 1, 1846 ; resigned January 1, 1867.

Dr. John W. Sawyer appointed January 1, 1867.

CONNECTICUT.

Retreat for the Insane, Hartford—April 1, 1824.

Dr. Eli Todd appointed January 7, 1823 ; died November 17, 1833.

Dr. Silas Fuller appointed June 11, 1834; resigned June 13, 1840.

Dr. Amariah Brigham appointed July 13, 1840 ; resigned August 16, 1842.

183

Dr. John S. Butler appointed May 13, 1843 ; resigned October
27, 1872.

Dr. James H. Denny appointed November 25, 1872 ; resigned
January 9, 1874.

Dr. Henry P. Stearns appointed January 23, 1874.

CONNECTICUT HOSPITAL FOR THE INSANE, MIDDLETOWN.

Dr. A. M. Shew appointed October 14, 1866.

NEW YORK.

BLOOMINGDALE ASYLUM, NEW YORK.

The Department for the Insane of the New York Hospital was
opened at the present site in June, 1821, under the name of
Bloomingdale Asylum, with Dr. James Eddy as Resident Physician and Dr. John Neilson as Visiting Physician. Dr. Neilson
occupied that position until January, 1831, when he resigned
and a Visiting Physician was dispensed with. Dr. Albert Smith
was Resident Physician from September, 1822, to March, 1824.
John Neilson, Jr., M. D., from March, 1824, to May, 1824.
Abraham V. Williams, M. D., from May, 1824, to June, 1825.
James Macdonald, M. D., from June, 1825, to December, 1830,
and Guy C. Bayley, M. D., from December 1830, till Dr. Macdonald returned from Europe in 1832. In May, 1831, Dr. James
Macdonald was appointed Physician, spent fifteen months in
Europe and returned in the latter part of 1832.

Dr. James Macdonald appointed October 13, 1832 ; resigned
August 15, 1837.

Dr. Benjamin Ogden appointed September 1, 1837 ; resigned
September 16, 1839.

Dr. William Wilson appointed September 16, 1839 ; resigned
April 1, 1844.

Dr. Pliny Earle appointed April 1, 1844 ; resigned April, 1849.

184

Dr. Charles H. Nichols appointed April, 1849; resigned May, 1852.

Dr. D. T. Brown appointed June, 1852; resigned January, 1877.

Dr. Charles H. Nichols appointed July 1, 1877.

NEW YORK CITY LUNATIC ASYLUM, BLACKWELL'S ISLAND,

Was in charge of Assistant Physicians from Bellevue Hospital until October, 1847, when Dr. M. H. Ranney was appointed Superintendent and he held the position till his death, December 7, 1864.

Dr. R. L. Parsons appointed January 7, 1865; resigned August 1, 1877.

Dr. Wm. W. Strew appointed October 1, 1877; removed November 19, 1879.

Dr. Thomas M. Franklin was appointed Medical Superintendent of the Branch Institution on Hart's Island on January 5, 1879 and transferred to the charge of the Main Hospital April 5, 1880.

KINGS CO. LUNATIC ASYLUM, FLATBUSH, L. I.

Dr. Francis Bullock, died in 1863.　⎫
Dr. Martin E. Winchell.　　　　　⎪　In the
Dr. T. M. Ingraham.　　　　　　　⎬　old
Dr. E. S. Blanchard.　　　　　　　⎭　Asylum.

Dr. Robert B. Baisely appointed October, 1855; resigned 1857.

Dr. John V. Lansing appointed May 7, 1857; resigned 1858.

Dr. Edward R. Chapin appointed May 7, 1858; resigned November 8, 1873.

Dr. Carlos F. MacDonald appointed November 8, 1873; resigned August 5, 1874.

Dr. James A. Blanchard appointed August 10, 1874; resigned May 23, 1877.

Dr. H. L. Bartlett appointed May 31, 1877, but the appointment reconsidered June 6, 1877, and Dr. Blanchard directed to remain in charge.

Dr. Blanchard removed July 11, 1877.

Dr. R. S. Parsons appointed July 11, 1877; removed August 31, 1878.
Dr. J. C. Shaw appointed September 1, 1878.

HOSPITAL FOR CHRONIC INSANE.

Dr. Guy D. Daly appointed January 1, 1877; removed July 16, 1881.
Dr. J. A. Arnold appointed July 16, 1881; transferred to Hospital October 15, 1881.
Dr. J. S. Woodside appointed October 15, 1881; removed November 30, 1883.
Institution placed in charge of Dr. Shaw December 1, 1883, and both buildings conducted as one Asylum.

NEW YORK CITY ASYLUM FOR THE INSANE, WARD'S ISLAND.

Dr. M. G. Echeverria appointed December 31, 1871; resigned November 6, 1872.
Dr. Theo. H. Kellogg appointed November 25, 1872; resigned August 1, 1874.
Dr. A. E. Macdonald appointed August 1, 1874.

NEW YORK STATE LUNATIC ASYLUM, UTICA—JANUARY 16, 1843.

Dr. Amariah Brigham appointed September 9, 1842; died September 8, 1849.
Dr. N. D. Benedict appointed November 3, 1849; resigned July 1, 1854.
Dr. John P. Gray appointed July 1, 1854.

MARSHALL INFIRMARY, TROY, NEW YORK.

Insane Department opened August, 1859.
Dr. J. D. Lomax appointed October 12, 1863.

HUDSON RIVER STATE HOSPITAL, POUGHKEEPSIE, NEW YORK.

Dr. J. M. Cleveland appointed March 28, 1867.

STATE HOMEOPATHIC ASYLUM FOR THE INSANE, MIDDLETOWN,
NEW YORK—MAY 7, 1874.

Dr. Henry R. Stiles appointed April, 1874; resigned February
9, 1877.
Dr. Selden H. Talcott appointed April 13, 1877.

WILLARD ASYLUM FOR THE INSANE, WILLARD, SENECA LAKE.

Dr. John B. Chapin appointed April 1, 1869; resigned August 2,
1884.
Dr. P. M. Wise appointed August 2, 1884.

STATE ASYLUM FOR INSANE CRIMINALS, AUBURN.

Dr. Edward Hall appointed November 6, 1858; removed June 1,
1862.
Dr. Charles E. Van Anden appointed June 1, 1862; resigned
February 17, 1870.
Dr. James W. Wilkie appointed February 17, 1870; died March
13, 1876.
Dr. Carlos F. MacDonald appointed March 17, 1876; resigned
October 1, 1879.
Dr. Theodore Deinen appointed October 1, 1879; removed May
15, 1881.
Dr. Carlos F. MacDonald appointed June 1, 1881.

BUFFALO STATE ASYLUM FOR THE INSANE—OPENED NOVEMBER
15, 1880.

Dr. J. B. Andrews appointed June 30, 1880.

BRIGHAM HALL, CANANDAIGUA.

Dr. George Cook, October 1, 1855; killed June 11, 1876.'
Dr. D. R. Burrell appointed November 14, 1876.

187

Sanford Hall, Flushing.

Dr. James Macdonald established an institution on Murray Hill in New York city in 1841, and removed to Sanford Hall in Flushing in 1845.
Dr. James Macdonald died in 1849.
Dr. Henry W. Buel in charge from 1849 to 1854.
Dr. J. Whitney Barstow took charge in 1854.

———

NEW JERSEY.

State Lunatic Asylum, Trenton.

Dr. H. A. Buttolph appointed April 19, 1847 ; resigned November 10, 1875, to take effect April 1, 1876.
Dr. John W. Ward appointed March 15, 1876; took charge April 1, 1876.

State Asylum for the Insane, Morris Plains.

Dr. H. A. Buttolph appointed June 29, 1875; opened the Institution August 17, 1876: retired from Superintendency January 1, 1885.

———

PENNSYLVANIA.

Pennsylvania Hospital for the Insane—Opened January 1, 1841.

Dr. Thomas S. Kirkbride appointed October 12, 1840 ; died December 16, 1883.
Dr. John B. Chapin took charge September 1, 1884.

188

FRIEND'S ASYLUM FOR THE INSANE, FRANKFORD.

Dr. Joshua H. Worthington appointed May 1, 1850; resigned
November 1, 1877.
Dr. John C. Hall appointed November 1, 1877.

INSANE DEPARTMENT OF THE PHILADELPHIA ALMSHOUSE.

Dr. N. D. Benedict appointed November 9, 1845 ; resigned
February 18, 1850.
Dr. Wm. S. Haines appointed February 18, 1850; resigned
February 11, 1853.
Dr. J. D. Stewart appointed February 11, 1853; died April,
1854.
Dr. A. B. Campbell appointed May 1, 1854; resigned July 2,
1855.
Dr. R. K. Smith appointed July 2, 1855 ; resigned July 21, 1856.
Dr. A. B. Campbell appointed July 21, 1856 ; resigned June 8,
1857.
Dr. James McClintock appointed June 8, 1857 ; resigned July 5,
1858.
Dr. R. K. Smith appointed July 5, 1858 : resigned September 24.
1859.
Dr. S. W. Butler appointed September 24, 1859 ; resigned De-
cember 1, 1866.
Dr. D. D. Richardson appointed December 1, 1866 ; resigned
September, 1880.
Dr. A. A. McDonald appointed September, 1880 ; resigned Jan-
uary, 1881.
Dr. D. D. Richardson appointed April 1, 1881.

Until 1859, the gentlemen above named were Medical Superintendents of the Philadelphia Almshouse, and the Insane Department was separated in 1859 when the office of Medical Superintendent of the whole Institution was abolished.

PENNSYLVANIA STATE LUNATIC HOSPITAL, HARRISBURG—OPENED
OCTOBER 1, 1851.

Dr. John Curwen appointed February 14, 1851 ; left February 12,
1881.

Dr. J. Z. Gerhard appointed December 30, 1880.

WESTERN PENNSYLVANIA HOSPITAL FOR THE INSANE, DIXMONT,
ALLEGHENY COUNTY.

Dr. Joseph A. Reed appointed April 1, 1857 ; died November 6,
1884.

Dr. Henry A. Hutchinson appointed January 15, 1885.

STATE HOSPITAL FOR THE INSANE, DANVILLE.

Dr. S. S. Schultz appointed May 21, 1868.

STATE HOSPITAL FOR THE INSANE, NORRISTOWN.

This institution was started with a double organization, a male
physician in charge of the wards for the men, and a female physician in charge of the wards for the women. It was opened
July 12, 1880.

Dr. R. H. Chase appointed May 20, 1880, to the charge of the
male wards.

Dr. Alice Bennett appointed May 20, 1880, to the charge of the
female wards.

STATE HOSPITAL FOR THE INSANE, WARREN—OPENED DECEM-
BER 1, 1880.

Dr. D. D. Richardson appointed January 21, 1880 ; resigned
July 1, 1881.

Dr. John Curwen appointed June 24, 1881.

BURN BRAE, KELLYVILLE.

Dr. R. A. Given, January 26, 1860.

MARYLAND.

MARYLAND HOSPITAL, BALTIMORE.

"The old Maryland Hospital was organized by Drs. Smith and Mackenzie in 1797. It was intended for a few lunatics, and as a general Hospital; as the city increased it was necessary to enlarge it, and it was extended in 1807, so as to accommodate a larger number of general patients and forty lunatics. At this time it was leased to Drs. Smith and Mackenzie, who attended all the patients and regulated every department. It was a private establishment until 1864, when the lease expired. The Legislature had appointed a board with power to appoint officers in 1828, which was done, and Dr. R. S. Steuart was appointed Medical Superintendent under the name of President, but did not until January, 1834, assume full authority. Dr. Steuart did not reside in the house, but visited daily for some time; but finding it necessary to have assistance, engaged five medical men to join in the medical administration of the house. This continued one year when a medical attendant was appointed, Dr. Wm. H. Stokes, who served one year. Dr. William Fisher became Resident Physician in 1836, remained four years, when his health failing he was relieved for eight months by Dr. H. Starr. Dr. Fisher continued until June, 1846, when Dr. John Fonerden was appointed."

Dr. John Fonerden died May 6, 1869.

Dr. William F. Steuart appointed June 9, 1869.

The patients were removed to the new hospital near Catonsville in July, 1872.

Dr. William F. Steuart resigned August, 1873.

Dr. John S. Conrad appointed August, 1873.

In 1876, the Hospital management was reorganized and a Board of nine members appointed by the Governor superseded the former self-perpetuated board with its Medical President who really acted as Superintendent. The office of Medical Superintendent replaced that of Resident Physician and

Dr. J. S. Conrad was appointed and assumed charge July 7, 1876 ; resigned March 14, 1878.
Dr. Richard Gundry appointed June 1, 1878.

Mt. Hope Retreat, Baltimore.

Dr. William H. Stokes appointed September 21, 1842.

DISTRICT OF COLUMBIA.

Government Hospital for the Insane—January 15, 1855.

Dr. Charles H. Nichols appointed October, 1852 ; resigned August, 1877.
Dr. W. W. Godding appointed September 1, 1877.

VIRGINIA.

Eastern Lunatic Asylum, Williamsburg—Sept. 14, 1773.

"Mr. James Galt was appointed keeper on September 14, 1773, and continued in office till the time of his death in 1801, and was suceeded by his son, William T. Galt, who continued in office till his death in 1826. Jesse Cole was appointed keeper in 1826 and resigned the same year ; was succeeded by Dickie Galt, who resigned January 1, 1837 ; succeeded by Henry Edloe, who resigned within a year after being appointed ; succeeded by Philip Barziza, who continued in office till July, 1881, at which time he was elected Steward of the Asylum, the functions of keeper and physician being henceforward united in one office."
Dr. John Siqueyra was appointed physician to the Asylum October 12, 1773, and resigned February 10, 1795, when Drs. Galt and Barrand were appointed physicians and continued in that

capacity until 1858, when Dr. Alexander Dickie Galt was appointed and continued till 1841, when he was succeeded by his son Dr. John Minson Galt, who died in 1862.

In charge of the United States Government from 1862 to the fall of 1865, when

Dr. Leonard Henley was appointed Physician.

Dr. R. M. Garrett was appointed February, 1866 ; removed 1867.

Dr. A. Peticolas appointed 1867 ; died November 28, 1868.

Dr. John Clopton acted as Superintendent until

Dr. D. R. Brower was appointed January 9, 1869 ; resigned November 17, 1875.

Dr. Harvey Black appointed November 17, 1875 ; removed March 9, 1882.

Dr. Richard A. Wise appointed March 9, 1882 ; removed April 16, 1884.

Dr. J. D. Moncure appointed April 16, 1884.

CENTRAL LUNATIC ASYLUM, RICHMOND.

Dr. David Burr Conrad appointed July, 1870; resigned November, 1873.

Dr. Randolph Barksdale appointed November, 1873 ; removed March 9, 1882.

Dr. Brooks succeeded him, but only lived about eighteen days after his election.

Dr. David F. May appointed April 11, 1882 ; removed April 23, 1884.

Dr. Randolph Barksdale appointed April 23, 1884.

WESTERN LUNATIC ASYLUM, STAUNTON—JULY, 1828.

Dr. Francis T. Stribling appointed Visiting Physician in May, 1836, and Superintendent and Physician in 1841 ; died July 23, 1874.

Dr. R. F. Baldwin appointed October 1, 1874; died November 14, 1879.

Dr. A. M. Fauntleroy appointed December 11, 1879 ; removed March 9, 1882.

Dr. R. S. Hamilton appointed March 9, 1882 ; removed April 15, 1884.

Dr. A. M. Fauntleroy re-appointed April 15, 1884.

WEST VIRGINIA.

HOSPITAL FOR THE INSANE, WESTON.

Dr. R. Hills appointed November 1, 1864 ; resigned July 1, 1871.

Dr. J. B. Camden appointed July 1, 1871 ; resigned May 16, 1881.

Dr. W. J. Bland appointed May 16, 1881.

NORTH CAROLINA.

ASYLUM FOR THE INSANE, RALEIGH.

Dr. Edward C. Fisher appointed Superintendent of Construction, September 15, 1853, and Superintendent and Physician in February, 1856 ; resigned July 7, 1868.

Dr. Eugene Grissom appointed July 7, 1868.

EASTERN NORTH CAROLINA INSANE ASYLUM, GOLDSBORO.

Dr. W. H. Moore appointed August 1, 1880 ; died December. 1881.

Dr. J. D. Roberts appointed January 1, 1882.

WESTERN NORTH CAROLINA INSANE ASYLUM, MORGANTOWN.

Dr. P. L. Murphy appointed December 9, 1882.

SOUTH CAROLINA.

LUNATIC ASYLUM, COLUMBIA—1827.

Dr. John W. Parker appointed December 24, 1836, previous to which the patients were attended by a visiting physician ; removed August 5, 1870.

Dr. J. T. Ensor appointed August 5, 1870 ; resigned January 1. 1878.

Dr. P. E. Griffin appointed January 1, 1878.

GEORGIA.

LUNATIC ASYLUM, MILLEDGEVILLE.

This Institution was opened for the reception of patients October 12, 1842. Dr. David Cooper was the first Superintendent and on account of feeble health resigned, and

Dr. Thomas F. Green was appointed January 1, 1846 ; died February 13, 1879.

Dr. J. O. Powell appointed February 13, 1879.

ALABAMA.

HOSPITAL FOR THE INSANE, TUSCALOOSA.

Dr. P. Brice appointed July 1860.

MISSISSIPPI..

LUNATIC ASYLUM, JACKSON.

Records of the Institution destroyed during the war.

Dr. W. S. Langley appointed in 1855 ; resigned in 1857.

Dr. W. B. Williamson appointed in 1857 ; resigned in 1859.

Dr. Robert Kells appointed in 1859 ; resigned in 1866.

Dr. A. B. Cabaniss appointed in 1866 ; resigned in 1869.

Dr. W. M. Deacon appointed in 1869 ; resigned in 1870.

Dr. Wm. M. Compton appointed in 1870 ; resigned May 3, 1878.

Dr. T. J. Mitchell appointed May 3, 1878.

LOUISIANA.

Lunatic Asylum, Jackson—Nov. 23, 1848.

William Collins was elected Superintendent and Dr. Selby Physician. Collins remained about nine months and in 1848 James King was elected in his place. In 1853, Edward C. Power was elected Superintendent in place of J. King. In 1854, Dr. Maybury was elected in place of Mr. Power. In 1856 James King was again elected. In 1856, Dr. J. D. Barkdull was elected and remained Superintendent until February, 1865, when he was deliberately shot and killed in the street in Jackson by a United States soldier. Mr. James King was again elected Superintendent and entered on the duties April 1, 1865. Dr. P. Pond was elected Physician in December, 1848, held the place about six years until Dr. Maybury was elected Superintendent ; was again elected in April, 1865.

Dr. Pond resigned.

Dr. L. A. Burgess.

Dr. J. Welch Jones, appointed March 17, 1874.

—

TEXAS.

Lunatic Asylum, Austin.

Dr. J. C. Perry was appointed Superintendent by Gov. E. M. Pease, who was Governor in May 27, 1857.

Dr. C. G. Keenan was appointed by Gov. H. R. Runnels in February 13, 1858, and during this time the erection of a portion of the present building was commenced.

" Sam Houston, who was elected Governor in 1859, appointed Dr. B. Graham in January 9, 1860, and under his supervision the present buildings and improvements were completed. About March 11, 1861, the institution was formally opened and during that month five or six patients were admitted. The secession

convention meeting about that time imposed such conditions upon the Governor and other State officers, as they were unwilling
to agree to, and Lieut. Gov. E. Clark, having assumed the gu-
bernatorial chair, reappointed Dr. C. G. Keenan, who retained
the position during Clark's term of office. F. Lubbock succeeding Clark as Governor in November, 1861, appointed J. M.
Steiner Superintendent November 1, 1861, who remained in
charge of the institution during the rebellion. August 21, 1865,
General A. J. Hamilton, who was appointed Military Governor
of Texas, appointed Dr. B. Graham September 9, 1865. But
upon another change of State officers in 1866, Gov. Throckmor-
ton appointed Dr. W. P. Beall Superintendent August 20, 1886.
Upon the re-establishment of a military government Aug. 1867,
Gov. Pease reappointed August 23, 1867,

Dr. B. Graham, who resigned March 27, 1870.

Dr. J. A. Corby appointed March 27, 1870; resigned March 1,
1871.

Dr. G. F. Weisselberg appointed March 1, 1871; resigned February 10, 1874.

Dr. D. R. Wallace appointed February 10, 1874; resigned April
18, 1879.

Dr. W. E. Saunders appointed April 18, 1879; resigned May 14,
1881.

Dr. L. J. Graham appointed May 14, 1881; resigned January
20, 1883.

Dr. A. N. Denton appointed January 20, 1883.

ARKANSAS.

State Lunatic Asylum, Little Rock.

Dr. C. C. Forbes appointed November 1, 1882.

TENNESSEE.

Hospital for the Insane, Nashville.

Dr. John S. McNairy appointed January 1, 1845; died August
18, 1849.

Dr. John S. Young appointed January 1, 1849 ; resigned March 1, 1852.

Dr. Wm. A. Cheatham appointed March 1, 1852 ; removed July 25, 1862.

Dr. Wm. P. Jones appointed July 25, 1862 ; resigned January 1, 1870.

Dr. J. H. Callender appointed January 1, 1870.

KENTUCKY.

EASTERN LUNATIC ASYLUM, LEXINGTON—1824.

Until 1844, the internal management and control of the patients was confided to a person known as head keeper, the prevailing idea being that the object of the asylum was in a great measure merely custodial.

Dr. J. R. Allen assumed the duties of First Superintendent on March 1, 1844, and held the office for a period of ten years.

Dr. William S. Chipley appointed April 1, 1855 ; resigned December, 1869.

Dr. John W. Whitney appointed December, 1869 ; resigned April, 1873.

The name of the Institution was changed in 1873 to First Kentucky Lunatic Asylum.

Dr. George Syng Bryant appointed April, 1873 ; died June, 1875.

Dr. R. C. Chenault appointed June 30, 1875 ; resigned May 1, 1880.

Dr. A. W. Bartlett appointed May 1, 1880 ; resigned February 1, 1881.

Dr. W. O. Bullock appointed February 1, 1881 ; resigned October 1, 1883.

Dr. R. C. Chenault appointed October 1, 1883.

Name changed to Eastern Kentucky Lunatic Asylum in 1876.

WESTERN LUNATIC ASYLUM, HOPKINSVILLE.

Dr. Samuel Annan was appointed by the Governor of Kentucky for four years from April 1, 1854.

198

Dr. F. G. Montgomery was elected by the Board of Trustees April 1, 1858.
Dr. James Rodman appointed June 1, 1863.

Third Kentucky Lunatc Asylum, Frankford.

Dr. E. H. Black appointed April, 1873.
The Institution was changed in February, 1874, and restored to its original purpose, an Institution for the education and treatment of feeble minded children.

Fourth Kentucky Lunatic Asylum, Anchorage.

Dr. C. C. Forbes appointed April, 1873.
Name changed in 1874 to Central Kentucky Lunatic Asylum.
Dr. C. C. Forbes resigned September 1, 1879.
Dr. R. H. Gale appointed September 16, 1879; resigned May 2, 1884.
Dr. H. K. Pusey appointed May 2, 1884.

OHIO.

Lunatic Asylum, Columbus.

Dr. Wm. M. Awl appointed May 21, 1838; resigned July 1, 1850.
Dr. S. Hanbury Smith appointed July 1, 1850; resigned July 1, 1852.
Dr. Elijah Kendrick appointed July 1, 1852; resigned July 1. 1854.
Dr. George E. Eels appointed July 1, 1854; resigned July 1, 1856.
Dr. R. Hills appointed July 1, 1856; resigned November 1, 1864.
Dr. Wm. L. Peck appointed November 1, 1864; resigned April 1874.
The institution was under the care of a building commission

from the date of Dr. Peck's resignation until the appointment of
Dr. Richard Gundry November 9, 1876 ; resigned May 16, 1878.

Dr. L. Firestone appointed May 16, 1878 ; resigned March 15,
1881.

Dr. H. C. Rutter appointed March 15, 1881 ; resigned November
22, 1883.

Dr. Thomas R. Potter appointed November 22, 1883 ; resigned
April 17, 1884.

Dr. C. M. Finch appointed April 23, 1884.

ASYLUM FOR THE INSANE, NEWBURGH—MARCH 5, 1855.

Dr. L. Firestone appointed December 1, 1854 ; resigned August
5, 1855.

Dr. R. C. Hopkins appointed May 8, 1855 ; resigned December
1, 1857.

Dr. Jacob Laisy appointed February 24, 1857 ; resigned October
11, 1857.

Dr. O. C. Kendrick appointed October 11, 1857 ; resigned Nov.
1, 1864.

Dr. W. M. Wythes appointed November 2, 1864 ; resigned Aug.
1, 1865.

Dr. Byron Stanton appointed Aug. 1, 1865 ; resigned May, 1869.

Dr. J. M. Lewis appointed May, 1869 ; resigned April 1, 1874.

Dr. Lewis Stuper appointed April 1, 1874 ; resigned August, 1875.

Dr. J. Strong appointed November 19, 1875.

ASYLUM FOR THE INSANE, DAYTON—SEPT. 1, 1855.

Dr. J. Clements appointed Sept. 1, 1855 ; resigned May 1, 1856.

Dr. J. J. McIlhenny appointed May 1, 1856 ; time expired May
1, 1862.

Dr. R. Gundry appointed April 15, 1862 ; resigned June 15, 1872.

Dr. H. B. Nunemacher acting Superintendent until Aug. 1, 1872.

Dr. S. J. F. Miller appointed August 1, 1872 ; resigned June 15,
1873.

Dr. H. C. Rutter acting Superintendent until March 1, 1874.

Dr. John H. Clark appointed March 1, 1874; resigned March 15, 1876.

Dr. J. R. Landfear appointed March 15, 1875; resigned April 15, 1878.

Dr. D. A. Morse appointed April 15, 1878; resigned July 15, 1880.

Dr. H. A. Tobey appointed July 15, 1880; resigned July 15, 1884.

Dr. C. W. King appointed July 15, 1884.

ASYLUM FOR THE INSANE, ATHENS.

Dr. Richard Gundry appointed June 16, 1872 ; resigned Dec. 19, 1876.

Dr. Thomas Blackstone acting Superintendent until January 17, 1877, when Dr. Charles L. Wilson was elected, but declared ineligible by the courts on account of non-residence, and Dr. Thomas Blackstone was again appointed acting Superintendent on March 20, 1877.

Dr. H. C. Rutter appointed March 30, 1877 ; resigned April 26, 1878.

Dr. P. H. Clarke appointed April 26, 1878 ; resigned April 19, 1879.

Dr. W. H. Holden appointed April 19, 1879 ; resigned May 6, 1880.

Dr. H. C. Rutter appointed May 6, 1880 ; resigned March 15, 1881.

Dr. A. B. Richardson appointed March 15, 1881.

CINCINNATI SANITARIUM, COLLEGE HILL, OHIO.

Dr. Wm. L. Peck appointed June, 1874; resigned August, 1875.

Dr. Wm. S. Chipley appointed August, 1875 ; died Feb. 11, 1880.

Dr. O. Everts appointed February 12, 1880.

LONGVIEW ASYLUM, CARTHAGE—JANUARY, 1860.

At Lick Run Asylum, Dr. J. J. Quinn was appointed in 1853 when the insane were removed from the Commercial Hospital.

Dr. Langdon was appointed in 1856, and Dr. Wm. Mount in 1859, who continued in charge until the removal of the insane to Longview Asylum in May, 1860, when the Lick Run Asylum was closed.

Dr. O. M. Langdon appointed November 10, 1859 ; resigned October 27, 1870.

Dr. W. H. Reynolds appointed December 14, 1870 ; resigned June 10, 1871.

Dr. Joseph T. Webb appointed June 10, 1871 ; resigned July 10, 1874.

Dr. Wm. H. Bunker appointed July 10, 1874 ; resigned April 17, 1878.

Dr. C. A. Miller appointed April 17, 1878.

INDIANA.

Hospital for the Insane, Indianapolis.

Dr. John Evans was appointed first Superintendent and continued in office about one year.

Dr. R. J. Patterson appointed July 1, 1848 ; resigned June 1, 1853.

Dr. James S. Athon appointed June 1, 1853 ; resigned Nov. 13, 1861.

Dr. J. H. Woodburn appointed October 31, 1861 ; resigned February 7, 1865. ·

Dr. Wilson Lockhart appointed February 7, 1865 ; resigned November 10, 1868.

Dr. Orpheus Everts appointed November 10, 1868 ; resigned June 7, 1879.

Dr. Joseph G. Rogers appointed June 7, 1879 ; resigned June 7, 1883.

Dr. W. B. Fletcher appointed June 7, 1883.

ILLINOIS.

Hospital for the Insane, Jacksonville.

Dr. J. M. Higgins appointed August 12, 1848; resigned June 9. 1853.

Dr. Andrew McFarland appointed June 16, 1854; resigned December 8, 1869.

Dr. H. F. Carriel appointed June 8, 1870.

Northern Illinois Hospital for the Insane, Elgin.

Dr. Edwin A. Kilbourne appointed September 14, 1871.

Southern Illinois Hospital for the Insane, Anna.

Dr. A. T. Barnes appointed September 23, 1873; resigned July 6, 1878.

Dr. H. Wardner appointed August 6, 1878.

Illinois Eastern Hospital for the Insane, Kankakee—Nov. 25, 1879.

Dr. R. S. Dewey appointed June 13, 1879.

Bellevue Place, Batavia—1867.

Dr. R. J. Patterson.

Oak Lawn Retreat, Jacksonville.

Dr. Andrew McFarland, May 10, 1872.

MICHIGAN.

Michigan Asylum for the Insane, Kalamazoo—April, 1859.

Dr. E. H. Van Deusen appointed October 19, 1855; resigned March 1, 1878.

Dr. Geo. C. Palmer appointed March 1, 1878.

203

EASTERN MICHIGAN ASYLUM, PONTIAC—AUGUST 1, 1878.

Dr. Henry M. Hurd appointed March 29, 1878.

WISCONSIN.

HOSPITAL FOR THE INSANE, MENDOTA.

Dr. J. Edwards Lee appointed June 22, 1859; resigned May, 1860.

Dr. J. R. Clement appointed May 22, 1860; resigned January 1, 1864.

Dr. A. H. Van Nostrand appointed April 20, 1864; resigned June 6, 1868.

Dr. Alexander S. McDill appointed June 1868; resigned October 1872.

Dr. Mark Ranney appointed April 29, 1873; resigned March 31. 1875.

Dr. Alexander S. McDill appointed April 1, 1875; died Nov. 12, 1875.

Dr. D. F. Broughton appointed January, 1876; resigned June 30, 1881.

Dr. R. M. Wigginton appointed July 1, 1881; transferred to Northern Hospital July 1, 1884.

Dr. S. B. Buckmaster appointed July 1, 1884.

NORTHERN HOSPITAL FOR THE INSANE, WINNEBAGO.

Dr. Walter Kempster appointed January 1, 1873; resigned July 1, 1884.

Dr. R. M. Wigginton appointed July 1, 1884.

IOWA.

HOSPITAL FOR THE INSANE, MT. PLEASANT—MARCH 1, 1861.

Dr. R. J. Patterson appointed July 6, 1860; resigned October 1, 1865.

Dr. Mark Ranney appointed October 1, 1865 ; resigned July 16, 1873.

Dr. H. M. Bassett appointed Dec. 4, 1873 ; resigned July 1, 1875.

Dr. Mark Ranney appointed July 1, 1873 ; died Jan. 31, 1882.

Dr. H. A. Gilmán appointed July 25, 1882.

HOSPITAL FOR THE INSANE, INDEPENDENCE—MAY 1, 1873.

Dr. Albert Reynolds appointed October 3, 1872 ; resigned Oct. 6, 1881.

Dr. Gershom H. Hill appointed October 6, 1881.

MISSOURI.

ASYLUM FOR THE INSANE NO. 1, FULTON.

Dr. T. R. H. Smith appointed April, 1851 ; resigned February 11, 1865.

Dr. Rufus Abbot appointed March 1, 1865 ; resigned March 1, 1867.

Dr. Charles H. Hughes appointed March 1, 1867 ; resigned May 29, 1872.

Dr. Thomas A. Howard appointed May 29, 1872 ; resigned Oct. 10, 1872.

Dr. T. R. H. Smith appointed December 22, 1872.

LUNATIC ASYLUM NO. 2, ST. JOSEPH'S.

Dr. George C. Catlett appointed 1874.

COUNTY LUNATIC ASYLUM, ST. LOUIS.

Dr. Turner R. H. Smith was first appointed October 4, 1867, but the order of his appointment was rescinded at the next meeting October 7, 1867.

Dr. Charles W. Stevens was appointed February 6, 1868, and re-appointed February 19, 1872.

Dr. Charles W. Stevens sent his resignation to the Court July 22, 1872, which was accepted July 25, 1872, to take effect August 15, 1872.

Dr. Turner R. H. Smith was appointed July 25, 1872, to take effect August 15. 1872.

Dr. Smith resigned January 6, 1873.

Dr. Wm. B. Hazard was appointed January 16, 1873; reappointed February 24. 1873; dismissed March 19, 1874.

Dr. H. S. Fichtenkamp was appointed Resident Physician Feb. 2. 1874.

Dr. Jerome K. Bauduy was appointed Visiting Physician Feb. 16, 1874, and resigned January 11, 1875.

Dr. H. S. Leffingwell was appointed Visiting Physician July 13, 1874.

Dr. N. DeV. Howard was appointed Resident Physician and Dr. E. S. Frazer Visiting Physician February 1, 1875.

Dr. N. DeV. Howard was reappointed Resident Physician and Dr. E. S. Frazer Visiting Physician January 31, 1876.

Dr. Charles W. Stevens was appointed Resident Physician Jan. 29, 1877.

Dr. N. DeV. Howard was reinstated by Mayor Overslotz and resigned finally in April, 1883.

Dr. Charles W. Stevens was appointed again and confirmed Nov. 6, 1883.

(From the Records of the City Register's office.)

MINNESOTA.

Hospital for the Insane, St. Peter.

Dr. Samuel E. Shantz appointed October 2, 1866; died August 22, 1868.

Dr. C. K. Bartlett appointed November 6, 1868.

Second Minnesota Hospital for the Insane, Rochester— Opened January 1, 1879.

Dr. Jacob E. Bowers appointed December 1, 1878.

NEBRASKA.

HOSPITAL FOR THE INSANE, LINCOLN.

Dr. Charles F. Stewart appointed November 15, 1871 ; resigned January 1, 1875.

Dr. David W. Scott appointed December 29, 1874; resigned February 27, 1875.

Dr. F. G. Fuller appointed February 27, 1875 ; resigned Nov. 16, 1877.

Dr. H. P. Mathewson appointed November 16, 1877.

——

KANSAS.

LUNATIC ASYLUM, OSAWATOMIE—NOV. 1, 1866.

Dr. C. O. Gause appointed May 11, 1866 ; resigned Nov. 30, 1871.

Dr. C. P. Lee appointed Nov. 1, 1871 ; resigned Nov. 30, 1872.

Dr. L. W. Jacobs appointed Nov. 1, 1872; resigned Oct. 1, 1873.

Dr. A. H. Knapp appointed October 1, 1873 ; resigned March 1, 1877.

Dr. F. B. West acting Superintendent from March 2, 1877, to November 1, 1877.

Dr. A. P. Tenney appointed Nov. 1, 1877 ; resigned Nov. 1, 1878.

Dr. A. H. Knapp appointed November 1, 1878.

INSANE ASYLUM, TOPEKA—JUNE 1, 1879.

Dr. B. D. Eastman appointed April 1, 1879 : resigned June 30, 1883.

Dr. A. P. Tenney appointed July 1, 1883.

CALIFORNIA.

INSANE ASYLUM, STOCKTON—JULY 1, 1853.

Dr. Robert R. Reed appointed July 1, 1853 ; resigned October 1, 1856.

Dr. Samuel Langdon appointed April 29, 1856 ; resigned August 1, 1857.

Dr. W. D. Aylett appointed August 13, 1857 ; resigned April 20, 1861.

Dr. W. P. Tilden appointed April 20, 1861 ; resigned August 1, 1865.

Dr. G. A. Shurtleff appointed August 1. 1865 ; resigned Sept. 1. 1883.

Dr. W. T. Browne appointed October 10, 1883.

STATE ASYLUM FOR THE INSANE, NAPA—NOV. 15, 1875.

Dr. E. T. Wilkins appointed March 16, 1876.

PACIFIC ASYLUM

Was established at Woodbridge, San Joaquim County, California, July 15, 1871, removed to Stockton September 15, 1877. Dr. Asa Clark.

———

OREGON.

HOSPITAL FOR THE INSANE, EAST PORTLAND, OREGON,

A private Institution caring for State patients.   Opened Dec. 1, 1862.

Dr. J. C. Hawthorne died February 1, 1881.

Dr. S. E. Josephi appointed March, 1881.

On the removal of the Insane to the State Hospital at Salem. October 23, 1883, the institution ceased to exist.

Oregon State Insane Asylum, Salem—Oct. 23, 1883.

Dr. H. Carpenter appointed September 1, 1883.

———

INSTITUTIONS IN THE BRITISH PROVINCES.

Asylum for the Insane, Toronto—Opened January, 1841.

Dr. Rees.
Dr. Telfer.
Dr. Parke.
Dr. Primrose.
Dr. Scott.
Dr. Joseph Workman appointed July 1, 1853; resigned July 19, 1875.
Dr. Daniel Clark appointed Nov. 18, 1875.

Asylum for the Insane, Kingston.

Dr. J. R. Litchfield appointed March, 1865; died Dec. 18, 1868.
Dr. J. R. Dickson appointed November, 1868; resigned December 31, 1878; died November 23, 1882.
Dr. W. G. Metcalf appointed July 1, 1879, having been Acting Superintendent during Dr. Dickson's illness from April 1, 1878, to the day of his resignation and then to the date of his appointment.

Asylum for the Insane, London.

The Asylum at Amherstburg was made separate from the Asylum at Toronto, September 25, 1861.
Dr. Andrew Fisher resigned June, 1868.
Dr. Henry Lander appointed June, 1868, and the patients were all removed to the new Asylum at London, November, 1870.
Dr. Lander died January 6, 1877.
Dr. R. M. Bucke appointed February 15, 1877.

ASYLUM FOR THE INSANE, HAMILTON.

Dr. R. M. Bucke appointed January 1, 1876. Transferred to London February 13, 1877.

Dr. J. M. Wallace appointed February 13, 1877.

QUEBEC LUNATIC ASYLUM, PROPRIETARY.

Drs. James Douglas, Joseph Morrin and Charles J. Fremont were its first founders and proprietors.

They fitted up for the cure and maintenance of the insane in 1845 a large building which they leased in the parish of Beauport. The present buildings were erected in the parish of St. Rock, on the road leading to the parish of Beauport, in 1848. Since then the establishment has been very much enlarged and a new one for male patients erected. The present proprietors are : ·

Dr. F. E. Roy.

Ph. Landry, M. P.

Dr. Ant. La Rue.

PROVINCIAL LUNATIC ASYLUM, ST. JOHN, NEW BRUNSWICK.

Opened by Dr. G. P. Peters December 12, 1848.

Dr. John Waddell appointed December 1, 1849 ; resigned Oct. 31, 1875.

Dr. J. T. Steeves appointed November 1, 1875.

NOVA SCOTIA HOSPITAL FOR THE INSANE, HALIFAX—OPENED DEC. 25, 1858.

Dr. James R. DeWolf appointed May 17, 1857 ; resigned April 1, 1878.

Dr. A. P. Reid appointed April 1, 1878.

HOSPITAL FOR THE INSANE, ST. JOHNS, NEWFOUNDLAND.

Dr. Henry H. Stabb appointed 1847.

HOSPITAL FOR THE INSANE, CHARLOTTETOWN, PRINCE EDWARD
ISLAND.

Dr. Mackieson appointed 1848 ; resigned 1874.
Dr. E. S. Blanchard appointed August 1874.

PROVINCIAL LUNATIC ASYLUM, ST. JOHNS, QUEBEC.

Dr. Henry Howard.

www.ingramcontent.com/pod-product-compliance
Lightning Source LLC
Chambersburg PA
CBHW030825270326
41928CB00007B/895